THE
BEAUTY
OF THE
METROPOLIS

AUGUST ENDELL

THE
BEAUTY
OF THE
METROPOLIS

*Translated from the German and
with an afterword by James J. Conway*

RIXDORF EDITIONS BERLIN 2018

The Beauty of the Metropolis by August Endell was first published in German as *Die Schönheit der großen Stadt* by Verlag von Strecker & Schröder in Stuttgart, 1908.

Articles by August Endell originally published in German in *Die Neue Gesellschaft*, Berlin, Volume 1: 'On Vision' as 'Vom Sehen' (no. 4, 26 April 1905), 'Evening Colours' as 'Abendfarben' (no. 7, 17 May 1905), 'Spring Trees' as 'Frühlingsbäume' (no. 8, 24 May 1905), 'Treetops' as 'Baumkronen' (no. 10, 7 June 1905), 'Potsdamer Platz in Berlin' as 'Der Potsdamer Platz in Berlin' (no. 12, 21 June 1905), 'The Art of Impressions' as 'Eindruckskunst' (no. 23, 6 September 1905).

This translation, afterword and anthology © 2018 James J. Conway

The Beauty of the Metropolis
first published Rixdorf Editions, Berlin, 2018

Design by Cara Schwartz

Printed by Totem.com.pl, Inowrocław, Poland

ISBN: 978-3-947325-06-1

All rights reserved. No part of this publication may be reproduced, stored in a retrieval system, or transmitted, in any form or by any means, without the prior written permission of Rixdorf Editions, except for legitimate review purposes.

RixdorfEditions.com

CONTENTS

THE BEAUTY OF THE METROPOLIS	7
Condemning the Age	9
Renouncing the Present	13
Love of the Here and Now	17
The City	23
The City as Design	27
The City as Working Entity	29
The City as Nature	35
The City as Sound	37
The City as Landscape	39
On Vision and the Visible World	41
The Scenic Beauty of the City	51
The Veils of Day	53
The Veils of Night	63
The Street as Living Entity	67
Notes	83
ARTICLES FOR *DIE NEUE GESELLSCHAFT*	89
On Vision	91
Evening Colours	97

Spring Trees	101
Treetops	107
Potsdamer Platz in Berlin	113
The Art of Impressions	119
Afterword	129

THE BEAUTY OF THE METROPOLIS

Then, with a hint of a smile, he would recollect the long streets of the metropolis. The metropolis, that lived on the blood, sweat and minds of the people, wolfed down its million-fold offerings and imprisoned all who streamed towards it, captured them forever within its granite walls, crushed them, stomped on them, annihilated them, roused the mad hopes of thousands, and let a few — at a high price — mount its battlements, hair already thinning, eyes and hearts numb. And despite all this, it was in the metropolis that the marvel of the present day emerged and where it dwelt.

Henning Berger[1]

CONDEMNING THE AGE

Since time immemorial, people have located a golden age in the past, old men have bemoaned the degeneracy of their time, the weary and the cowardly have cloaked their incompetence by complaining bitterly about the depravity of the age. But no-one ever took them particularly seriously. Only now, it seems, are these complaints becoming louder and more insistent, indeed they exert an alarming influence over our lives. To revile the present day is almost a reflex; people rebuke the degeneration, the anxiety, the haste of our era, and with fine naivety find a parallel between the speed of the automobile and the pace of life. Good people who know precious little of the past and nothing at all of the maladies of the mind hold our age accountable for every madman's deed, with harsh words and no further ado. And when a minister openly declares that our present-day lives are entirely lacking in ideals, that only material interests prevail, that base egotism, hedonism and avarice alone are the driving forces of our actions, there is no-one to stand up and refute such nonsense, instead all 'serious' men nod their heads in grave agreement.

It is an order of judgement completely and utterly fatuous in itself, for you can criticise anything with a lifeless, lying morality. Of course material interests prevail today – they always have. The objective of humanity, whether it accepts it or not, is acquisition. And every ideal the world has ever created originated in this gainful existence. All the virtues we know have just one purpose for our acquisitive species. Even those that might seem least earthbound – maternal love, goodness, tenderness, sympathy – only gain meaning in the hard realm of necessity in which we dwell (to the extent that they are virtues at all and not positive experiences for those that practise them). And in our world, these virtues of sympathy, goodness and love are only blessings for their beneficiaries when conjoined with earnestness, restraint and relentless determination. Necessity prevails, and it forces us to be selfish. Spare me the pious lament that money is the measure of all things. Money is merely a symbol of the necessity that circumscribes our lives, compelling us to action, to a full life, impersonal, oblivious to human considerations, hard yet never evil. And there could be no gift more demonic than the dream of the utopians – a secure livelihood for all. Our whole wondrous world would collapse into ruins. Everything springs from necessity, the 'base' drive to acquire. Even art, which is now so often subject to bombastic, sanctimonious glorification. It is only the hard steel of life that has nurtured the passionate energy of artistic creation in frivolous and idle humanity. Every ideal is born of earthly endeavour, its meaning and value accorded on our Earth alone, in our activity; but any ideal that tries to reach beyond this world is nothing more than a cloak with which the faint-hearted conceal their fear of necessity.

Life would be utterly impossible without ideals. Search and you will find them everywhere in the working life of our time. Fertile invention, dogged endurance, admirable consistency, shrewd planning and the most diligent individual works create the whole bright and lively world around us, in an age greater and more industrious than any before. We have every reason to rejoice when we see this working culture developing before our eyes.

On one point, however, the accusations are just – as strong and vibrant as this working culture is, it lacks design, it lacks a culture of pleasure. We excel at rolling iron, spinning yarn and constructing machines. But when it comes to erecting buildings or forging trellises with that iron, making fabric from that yarn, making utensils with that machine, our creative powers fail us. Seeking guidance for our forms we anxiously look to foreign cultures, and to the past.

RENOUNCING THE PRESENT

We are ashamed of our wants and needs. But the curious thing is that the blame lies entirely with those who loudly denounce our age and preach renunciation of the present – all manner of Romantics who proclaim that our only resort is to escape into nature, into art, into the past. It is they who have misled the worker, bringing conflict and uncertainty to his life. They have convinced him that his actions are a mournful necessity at best, and that he should reserve 'ideals' for his leisure time at least. And in so doing they have fragmented life, severing the link between work and leisure; the wondrous refrain of our working lives is stilled in those places where we pursue our pleasures, where we shape our surroundings. We must make do with surrogates, imitations, a bastardised antiquity. Our lives are robbed of unity, our desire – which springs from our entirely healthy working lives – numbed by empty words, mawkish phrases and vacuous fabrications, robbing us of the greatest good: the sincerity and purity of sensation.

There are those who arrogantly reproach the present day, the very culture itself, and preach a return to nature. As if the 'nature' these shallow people know weren't exclusively formed by human hand. As if all culture, all human endeavour were not also nature. They wax sentimental over the polity of the bees – which are otherwise indistinguishable from wasps, their sting feared for no reason – but the much finer, more rarefied structures of human organisation seem unworthy of admiration. The word 'nature' becomes a battle cry to cover every form of stupidity. These 'nature-lovers' don't even bother to understand what they claim to love. They see in it only what others before them have seen in it. They see it through the prism of literature. And they're filled with moral indignation when painters discover and dare to depict a beauty in it that the eyes of the fool can't even make out after the fact. They love nature because it doesn't affect them, because they want nothing from it, certainly nothing that might be attained by tiring struggle or passionate longing. So this return to nature becomes an escape into a vacuous, artificial fantasy world, contrived by weakness and fear, offering neither truth nor health nor redemption.

They escape into art! Not to the art of the living – which has nothing to offer the fugitive from life – rather to a theatrical, tarted-up art, a transcendent art, which bears no relation to reality. They worship art, they garland it with a gleaming halo of supernatural effect and raise it to a religion – you can't escape surrogates anywhere, it seems – and this senseless adoration demeans it more than the most appalling barbarism. Art is a craft, a profession like any other, its practical value derives from its effect on the mind, but it does not paint a better world or paper-thin ideals. That it continually testifies to the wonders of this world anew should suffice. It echoes the rhythm and refrain in the life of the worker. But no-one wants to hear it. Exaggerated worship requires ascension to the infinite, weakness and prudery call for mawkish modesty, ponderous patriotism needs vainglorious heroism. This is why they shun the new, emerging art in favour of escape to the old and venerable, which seems to represent an existence that is unfamiliar and thus superior; they try to steal its beauty, misappropriating it for undignified farce.

They escape to the past! Not that the nostalgist really understands history, or truly honours it. They contrive a strange, distorted world from meagre schoolroom knowledge and dutiful visits to the theatre, a tame, agreeable world with no thorns, no obstacles, no suffering, but plenty of heart-warming sensation. Nowadays you have to pay rent and work in an office but in Athens you would have been Pericles himself, or Phidias, even, and in Florence you wouldn't fester away on the fifth floor, oh no – you would be on intimate terms with the Medici. Back then everything was different, back then the diligent, the brave, the genteel still counted for something. In any age you might have amounted to something, it's only this one that fails to recognise virtue. But the nostalgists don't realise that the ruins of the past only come alive to those who understand the present, and that there is no age we can understand as well as our own. They don't realise that the most essential things in life remain unchanged by the passing of time, that human struggle, hardships and joy, that the sum of life was ever thus, that only fools and weaklings seek a golden age either before or after their own. And it is only those who fearfully withdraw from necessity, trying to ignore their obligations in a world of illusion, who are denied joy.

Romanticism is the mortal enemy of the living. It makes the inept arrogant, it confuses and inhibits the worker, it falsifies sensations and feelings, breeds insincerity and sentimentality, it is an enticement to empty masquerades, to resounding speeches that deceive the masses, destroy the unity of existence, sever the connection between longing and life, ideal and deed. It views the living with suspicion, poisons people's confidence in themselves and undermines unified, simple, straightforward, self-assured action.

LOVE OF THE HERE AND NOW

To any culture there is but one sound basis – the passionate love of the here and now, of the present day, of the land in which we live. You hear so much talk of patriotism. Germany extravagantly claims its superiority in every field – German thinkers, German poets[2], German painters are elevated above all others, any foreign influence denied or angrily contested with strange zeal. As if that's what it came down to, as if we loved our mother because she were wisest and fairest among women and not because we are bound to her by thousands upon thousands of enduring bonds, because she knows us better, more completely and more intimately than anyone else. What on earth will our poor patriot do when he realises one day that Shakespeare is greater than Goethe, that Romanesque architecture, which they call German these days, originated in Rome and Syria, or that the French too have temperament[3] – even if they don't share our word for it? When will we see an end to this ridiculous arrogance that condemns whole peoples while pompously placing itself at the summit of all mortals? Few of our writers can compare in heart and purity with the singular Balzac. Admittedly, there are many among us who seem to regard the heart as an item of clothing to be worn as conspicuously as possible. But perhaps we may also learn that boldly proclaiming your own temperament isn't the same as actually possessing temperament, or heart. Would we love our country any less were we no longer able to boast that we alone possess temperament? Can such things even be calculated, or measured? Who is to say whether Kant and Bach are greater than Dante and Michelangelo? Even if our country were a paltry pauper in every regard, with no glittering past, even if it had never sired genius – would it not still be our *Heimat*[4]?

Those who can only justify their love of fatherland through reckoning, by arrogantly condemning the lives and works of foreigners, have no idea what *Heimat* is. *Heimat* is not, as is widely assumed, synonymous with the land in which you live, but rather the land with all its life in a certain state, experienced at a certain time and informed by an entirely personal perspective. Each of us creates our own *Heimat*. And what applies to the individual applies equally to an entire people – its *Heimat* is composed of the shared fragments of individual worlds; it is that which the collectivity experiences as environment, as the shared substratum of its life. It is not the soil or the form that it has assumed over time, the landscape and the cities as they are, but rather how they are perceived and experienced. This is why the same environment may give very different senses of *Heimat* to different people. The more intensely we experience the land and the age, the richer, broader and more distinctive our sense of *Heimat* becomes. This is why *Heimat* isn't something fixed or immutable, rather it is in a state of becoming, constantly changing, determined by our lives and above all our perspectives. It is something completely unique to each person, that cannot be compared to anything else, and for this reason we cannot assess its value against the *Heimat* of another, no more than we can assess the world. And when pessimists insist on doing so all the same, they fall victim to a fallacy we may expect and pardon in the naïve, but which the thinking person should never entertain. To those struggling in life, every setback seems like a hard failure of fate, and it is understandable that they may long for unbroken success while utterly failing to notice that only failure lends strength and value to our desires, that good and bad luck are indivisible, because it is only out of their coexistence that we derive the direction and momentum of our lives. It is only the constricting banks that form water into a current, only necessity compels easily fatigued, frivolous humanity to forward-thinking, well-considered action. It is worry and torment that engender happiness; without them we would never experience it as such. And that which may be a matter of concern or misfortune in the life of the individual becomes an abyss of vice and squalid distress in the life of a people. And they pass by with a pious shudder, denouncing the depravity of the age. This complaint is nothing but a fear of life – Romanticism. How tawdry – imagining yourself to be pure just because you avoid dark things. Here, too, we are confronted with new challenges. Here, too, necessity is the mother of action. And in the present day we can act, while in the face of past suffering our hands are tied. There are certain moments in earlier eras that are impossible to dwell on without an overwhelming sense of torment and helplessness. But in the present day even the most terrible things may inspire us to action.

A life without suffering, a *Heimat* without suffering – these are things for which only a child would wish, or those who do not realise that life is movement, action, fulfilment. The productive among us do not evaluate *Heimat* by the petty calculation of what it might provide us, but instead strive to experience it. Its bright, dynamic nature is revealed only to those who immerse themselves in it. You have to know how to seize it. *Heimat* isn't a lifeless gift that you receive whether you want it or not. In truth it is accessible only to those who seek it out. Only that which we experience is *Heimat*. For this reason alone, establishing a national heritage on the achievements of our ancestors is foolish. Bach and Kant belong to those who comprehend and experience them, not to those who happen to be born in the same region.

Heimat must be attained. It arises only in the eye of the seeker. It is those who refuse to seek it who condemn the here and now, taking refuge in a distorted dream world; in their ignorance they regard reality as too small, too meagre; mean and embittered, they forego treasures immeasurable. The world is endless, not just in space and time, in great and small, but also in the different ways in which we can behold it. Time after time humanity has believed itself to have arrived at its outer limit, expecting nothing more, and time after time, just a little more searching has opened one new door after another. The anxious temperament, pious and humble, warn us not to search too far, not to chase down the secrets of the unknown, because we might destroy the joys of this world entirely. What infantile blasphemy! As if it weren't the most magnificent certainty that the world never and nowhere comes to an end, that we may lift every veil with impunity, each one revealing new wonders, new secrets. Seeking has only ever enriched, has only ever revealed beauty unseen. Admittedly not everything is beautiful. You have to be selective. Nature doesn't ask after our desires, it has its own ways and means. But the most astonishing thing is that despite this, everywhere – no matter where we look – we are offered fulfilment in endless abundance.

That is why attaining *Heimat* is such an inexpressible joy, such limitless enrichment for the soul. Those who open their eyes, who seek without prejudice or petty motives, who can devote themselves utterly to the here and now, will experience *Heimat* as a miracle, brighter and richer than any dream. It will become a mother offering endless renewal. Because there is joy in observing and absorbing, there is reinforcement, redemption and rejuvenation. Because observation and act are one, not some narcotic dream that leads off into the distance before depositing the sleeper back in waking life with a sudden jolt, rather an experience in broad daylight that inspires us to live in the here and now.

This is why this observation of *Heimat* is the

only sure guide towards a unified culture, to design. We can only shape the present, and this present can only be shaped by those truly able to experience it with all its opportunities, its means of development, its requirements. Knowledge of foreign or past desires is of no use here. The Gothic marketplace and the Baroque square – these are masquerades now. Only those who feel the rhythm of today's city will be capable of building the city that we need.

Above all, we need immersion in *Heimat*, knowledge of its essence, a vital sense of its dynamism. Liberation from all those myopic, mawkish phrases that have debased the word *Heimat* and robbed it of all strength of meaning. Genuine love for the fatherland, passionate love of the here and now. This path is not easy to discern, because boorish patriots have so often sullied that which is most precious, and in so doing reveal their contempt. The only guide here is unprejudiced, intensive observation. A unified national sense for these things will only arise if we continually attempt to express what we experience.

In the modest offering that follows I shall make this attempt.

THE CITY

Of course the metropolis, the most visible and perhaps most singular incarnation of our present-day lives, the most conspicuous, most complete form of our works and desires, has always been the subject of disproportionate attack. The city appears as a symbol, as the most potent expression of a culture that has turned its back on the natural, the simple, and the naive, a heinous chaos overflowing with things calculated to disgust the upright citizen – desolate hedonism, nervous haste and sickly degeneration. It draws in newcomers with false enticements and then spoils them, enervates them, turns them feeble, selfish and mean. City-dwellers are mocked for their lack of *Heimat*. People complain of the unspeakable ugliness of the city, with its chaotic din, its dirt, its dismal courtyards and its thick, murky air. It would be easy to ignore these opinions, like so many others, were it not that city-dwellers themselves wish to believe it – imagining *Heimat* to be the little peasant cottage with the window shimmering in the setting sun so familiar to them from the theatre – and if thousands of people hadn't needlessly stunted their own lives with such talk. You might well consider it a worthwhile objective for every city to disappear from the face of the Earth. But for now they do exist, and must exist if our entire economy is not to collapse entirely. Hundreds of thousands must live in the city and instead of instilling in them an unhealthy, hopeless longing, it would be a better idea to teach them how to really see their city and to draw as much joy and strength from their environment as possible, comparatively little though that may be in absolute terms. We may readily admit that life in the city is more stressful, less healthy than it is in a little village in the country. We may complain that city-dwellers are ever more estranged from the soil, from plant life, from animals, and that this robs them of numerous opportunities for happiness. We must also concede that most of our structures are hopelessly dull and lifeless yet somehow showy and pretentious at the same time. But this gives rise to the challenge of redesigning our cities to make them more spacious, more respectable, more artistic, as well as the other, more readily achievable task of compensating for these shortcomings with other pleasures.

Because this is the amazing thing – despite all of its ugly structures, despite the noise, despite everything in it with which you might find fault, the city is a miracle of beauty and poetry to those who see it, a fairy tale, brighter, more colourful, more varied than anything the poet might relate; a *Heimat*, a mother who every day showers her children with fresh joy in boundless extravagance.

That might sound paradoxical, or exaggerated. But when we resist blind prejudice, when we learn how to surrender ourselves, when we concentrate on the city, observantly and insistently, we soon become aware that its streets in fact contain a thousand beauties, wonders without number, riches without end, right before our very eyes and yet seen by so few.

We look to the cities of the past in stunned admiration – Babylon, Thebes, Athens, Rome, Baghdad. They all lie in ruins and even the most industrious imagination is powerless to reconstruct them. But our cities are alive, they envelop us with the entire force of the present, of existence, of the now. And any tradition, even the most precious ruins, are lifeless, ghostly and meagre compared to its bright infinitude. Our cities are as inexhaustible as life itself, they are our *Heimat*, because every day they speak to us in a thousand voices that we can never forget. However we view them, they give us joy, they give us strength, they give us the soil without which we cannot live.

THE CITY AS DESIGN

Admittedly there is one area in which our cities compare poorly to those of antiquity: they have no form, no design. Our streets have grown wider, our buildings taller and more expansive, but we haven't managed to breathe life into the raw forms dictated by economic and technical requirements. Our streets have no essence of their own, no manner or character unique to them. Our squares are empty spaces devoid of grandeur or form, buildings don't blend in with streets, they somehow manage to be loud and obtrusive without making an impact. There's no cohesion between structure and thoroughfare. And this is lamentable, but hardly surprising when you consider that industry, technology and trade have monopolised all the dynamic, creative talent in recent decades, and that only now, as new creations abate in those areas, are forces being freed up for artistic design, forces that are slowly embarking on the conscious design of what had previously accrued by chance and blind necessity, without care, without love.

THE CITY AS WORKING ENTITY

But for all this, the city is even more beautiful as a working entity, a functioning construct. Just like nature, work in itself is entirely devoid of any beautifying intent or purpose. Work aims solely at acquisition and as such we initially experience it as an onerous burden, a hardship, an irritant. But to the observant, work is full of the most varied beauty, like any natural entity. Beauty can be found in even the lowliest labour. Unfortunately this is only apparent to the labourer himself, and even then not always. It is a beauty often imperceptible to the senses, a beauty only apparent in the conception, in the idea, like the beauty of a mathematical proof that lies not in its result but in the rhythm of its working, like the beauty of an ingeniously conceived and implemented experiment, the beauty of a scientific demonstration. Sadly such things are inaccessible to most. Experts may experience them subconsciously, because the power of this beauty underpins and maintains their lives, but they are not always aware of it; knowing nothing of its power, they are unable to communicate it to others. It is truly a shame that this working beauty lies secretly stockpiled by the sciences, completely unused, that the almost caste-like nature of the sciences remains completely closed to the amateur, while wretched popularisation – even more obnoxious when rendered 'poetic' – maps out an illusory image of no value whatsoever. Anyone who imagines science must be made digestible to the amateur with wretched gimmicks knows nothing of its beauty.

The same goes for industry and manual labour. How much beauty is there in the labour concealed in the workshops and factories of the city? How many cleverly fashioned levers, devices and machines are there, how ingeniously do the different stages of work interlock? An abundance of ingenuity, imagination, wisdom and consistency wherever you look. Should you be disturbed by the chaotic noise of a neighbouring factory, call by and have them demonstrate the work stages, explain the machines. You will be amazed to discover the wonderful conceptual structure concealed beneath the noise, dust and dirt, and you will find meaning in this noise; you will learn to differentiate the voices of the machine, the chaotic din will become a language. How strange the different work stages are in themselves, what a remarkable entity the machine is when you finally comprehend its inner being, the working of its parts. The walls of the cylinder, meeting the constant rise and fall of pressure, the valves and their relentless to and fro, the base plate bound to the foundations by giant bolts, its heavy bulk easily absorbing all the jolts and vibrations – a whole realm of forces and internal motion. Then there's the iron bridge, composed of hundreds of working elements, straining with all its force, subtly stretching and then elastically contracting with each new burden, the main parts moving against one another, the whole thing efficiently shifting in steel joints along roller carriages as it absorbs strain, sun, cold, in a gentle, barely perceptible oscillation of dilation and contraction. There's a strange charm that comes from discovering empathy for the secret life of this behemoth. Admittedly, comprehending it all requires at least a little understanding of the structure and construction of such things. That's why we should keep on discussing this beauty of labour, and why experts should explain their work in comprehensible language – for everything, even the most difficult things, can be expressed simply and comprehensibly. Experts will delight in the novelty of viewing their own activities as an artistic undertaking, while outsiders will become aware that there are a thousand other worlds alongside their own. And while none of this will lighten life's burden, it will surely supply more strength to bear it.

And perhaps these technical formations are not even the greatest things that lie concealed in the working life of the city. The cohabitation of large amounts of people inevitably results in thousands of regulatory arrangements. Like the thick swarm of bees unintentionally creating hexagonal cells between the warmth of their forms, densely packed city-dwellers give rise to remarkable formations that wrest order from association, cohabitation and mutual dependence. It would be tempting to depict the ways in which necessity imposes order on their relations, and what beauty there is in that order. The organisation of a large business, its structure, its construction, its operations, its accounts, its systems of control – these alone are remarkable enough. As clean, pure, consistent and transparent in their forms as crystals. Necessity cannot abide the superfluous, and it compels us to clarity. More varied and richer still are the organisation of the state and its cities; here too there is coherence, constant renewal and development that can only be compared to natural formations. Admittedly we tend to perceive this as nothing more than a necessary evil, and naturally it is most apparent at the points of friction; it is precisely what doesn't work, the things with which we struggle that become apparent in the little political battles of the day, and these things should be eliminated. It is the irritants that come most readily to mind, and the vast scope of these arrangements means that only a few are accorded an overview, and even they are not always aware of the beauty of their forms. And rarer still are the lawyers or administrative professionals who take it upon themselves to offer the amateur an insight into these unusually rich formations.

And so we don't see nearly enough of this beauty of labour – an important element in every individual's life, but not sufficiently tangible for the amateur, visible only to those not disheartened by the effort of making their way through dry and arcane nomenclature, through obscure descriptions, to arrive at a view of this world. Simplified depictions could reveal endless riches here. There could hardly be a more valuable service than making the working life of individual professions comprehensible, making their beauty vital and tangible.

This knowledge would bring unity to our culture, context to our work, impetus and enthusiasm to our collective endeavour.

THE CITY AS NATURE

As well as this hidden beauty, the one that speaks not to the senses, accessible only to those who have the perception, the reason and the imagination required to sensitively investigate work formations, there is also the beauty of the city as nature. This may sound odd, precisely because this beauty is almost always overlooked, because we are not used to regarding a city as we might regard nature – the forest, the mountains and the sea.

THE CITY AS SOUND

How strange that the cawing of the raven, the wafting of the winds, the roaring of the sea should strike us as poetic, grand and noble. But the noises of the city are evidently not even worthy of our attention, and yet they in themselves form such a remarkable world that even the blind can experience the organism of the city in all its abundant variety. Just for once, listen, listen to the voices of the city. The light rolling of the cab, the heavy rumble of the post coach, the clacking of hooves on the asphalt, the quick, sharp staccato of the trotters, the dragging tread of the cab nag, each with its own particular character offering more subtle gradations than word can tell. We can differentiate one vehicle from another without really knowing how, and without relying on our eyes. The noises are as familiar to us as old acquaintances. Admittedly they are often too loud, even deafening in close proximity, but almost always attractive as they move off and the noise gradually dies away, becoming softer in the distance. How cheerful the sound of the rolling wheels, how wonderful the sudden hush when a coach disappears into a side street. How insistent the stomping gait of a lone pedestrian. How fleeting, soft, almost tender the sound of people treading a narrow street rarely bothered by vehicles – the kind of thing you can often hear in Dresden's Schloßstrasse[5], for instance. What muffled passion in the shuffle and shift of waiting crowds. How varied the voices of the automobile – the buzz as it draws near, the shriek of the horn, and then, gradually more audible, the rhythm of the striking cylinder, its regular metallic beat alternately hissing, thrusting and subdued. And finally, the siren song of the wheels close by, their spokes striking the air, the soft sliding grind of the rubber tyres. How mysterious the deep humming of the transformers hidden in the advertising columns[6], barely audible, reaching us like a dog gently nudging its master from behind. How wondrous the plump, dark roar of a tram at full speed, the heavy stamping of the carriage marking time, gradually joined by hard taps against the tracks, the jangling of the wheelworks, the whirring of the pulley and the long resonant hiss of the wire. You can wander through the city for hours, listening to its voices soft and loud and experiencing the great connections of life, in the silence of secluded areas and the roar of busy streets. The charm of it all is beyond words.

THE CITY AS LANDSCAPE

As vivid and richly differentiated as the metropolis is for the listener, to the observer it is an inexhaustible gift. The landscape of the city offers a colourful, constantly changing spectacle, an abundance, a wealth that the generations will never exhaust. And if this is now apparent to only a few, it is simply because we humans can only arrive at our pleasures through struggle – the beauty familiar to us now was only discovered gradually over the course of millennia, eventually forming an immutable cultural heritage. The metropolis is still so young that its beauty is only now being discovered. And here, as with any cultural bequest, every new form of beauty seems strange at first, encountering suspicion if not outright abuse. The age that brought the great expansion of the city also sired the poets and the painters who began to sense its beauty, and to turn this beauty into art. But people deluged them with suspicion, invective and moral indignation, damning the artists who dared descend to the dirt of the street, little realising what honour this judgement bestowed. This is precisely why these artists will endure – in the dirt of the street that crowds pass by with a contemptuous shudder, in the crush of the most egregious selfishness, the most heedless acquisition, they found beauty, grandeur and tenderness. And because this discovery and this pursuit remain misunderstood, angrily dismissed as depravity and treason, I wish to expand on this beauty, especially as the vocabulary of the visible is richer, this world lending itself to greater description than that of the merely audible.

ON VISION AND
THE VISIBLE WORLD[7]

But first I would like to resolve a series of misunderstandings that often seem to present an impossible impediment to those who might otherwise find their way to this beauty. It is naive to assume that one person sees the same way as another, and that if ten people observe a thing, their ten entirely divergent impressions can be solely attributed to their differing tastes. The assumption is that the images of this thing can proceed whole and unhindered through the retina and into the mind. And this delusion, which has a dubious enough influence in epistemology, is perhaps even more harmful to aesthetic observation. Certainly the retina receives a complete image of the object, but the sensitivity of the retina varies; it is strongest at the macula lutea[8], weakening towards its perimeter. So to see an object completely, we must apply the macula lutea to all of its parts. This can take some time, especially if we're capturing three-dimensional objects that require us to constantly change position. Life would be impossible if we attempted to examine everything in detail. This means that people generally don't view things with great precision. Moreover, it is not the perceptual image in itself that interests us, but the object, which is something different and only transformed into a perceptual image by the mind. Initially the only things that interest me about a table is whether it has a surface at a certain height above the ground, if it has legs that I might bump into, and perhaps whether it has a drawer. Ascertaining these things requires nothing more than a fleeting glance and a few probing taps of hand and foot. A knife only interests me for its handle and its cutting edge, and so on. In short, in our practical lives our vision serves only for orientation, from the wide scope of the visible we see only that which is of significance to the course of our desires, everything else remains a vague impression, soon to be forgotten. Even those pieces and parts that we examine more closely are only partially recorded. Everyone knows exactly where to find the front door of their building, where the door handle and doorbell are, but the individual construction of the door, its colour, the shape of its glass panes – all of that remains relatively unseen. Children spend a good part of their early years liberating themselves from a surfeit of visual impressions, and reducing them to the most essential points.

Consequently it took a long time for people to discover that vision could be a pleasure in itself. Humans started reproducing the visible early on, but for a long time this was only to record the figures depicted, because they loved the living thing – the hero, the horse – not because they thought that the form of the horse was beautiful in itself, that it might bring joy. Naturally there had always been an element of visual pleasure in the love of the thing. But only when it was depicted did they begin to sense that the form itself might be something wondrous, distinct from the object being depicted. And this has only become clearer over the course of the centuries – to artists, initially, of course. But only in our present era have we slowly begun to realise that form and colour do not derive their beauty from the object, that there may indeed be a beauty that is not even perceived in the object when we observe it for purely practical purposes, and that it is only artistic vision that confers upon the object the beauty that resides in form and colour, independent of all material relationships. Seeing a painting of Bismarck we recollect the man himself, and there is a joy that accompanies this recollection, but there is also joy in the form and colour, in the head, the mouth and the eyes, that remain accessible to those who know nothing of Bismarck. And it is only this joy that is artistic. The inept will always invoke representational relations to mask their inability to impart beauty. But this is a deception, and one that cannot hold. Generally, as time passes, so too does interest in the object, while purely visual beauty remains undimmed. The doges – their struggles, their wisdom and their grandeur – are almost completely forgotten, difficult to reconstruct with our mind's eye, but the Doges' Palace is no less alive than when it was first built.

So in life the image of the face has a purely functional role; it is only artists who made it the focal point of their work and recognised its value for our emotions. By reproducing it, they brought humanity tidings from a second world alongside and amid the world of objects – the world of the visible.

Unfortunately this relationship became utterly confused when people attempted to explain modern painting by claiming that it is not concerned with the 'what', but the 'how', that the technique, the method of painting is the central issue, and that only this lends value to a painting. This was an attempt to express a genuine relationship but this alluring antithesis was too oblique and ruined everything, making it harder than ever for the outsider to access this new vision. It is not the wonderfully accomplished technique with which Manet painted asparagus[9] that concerns us, rather his revelation that a bundle of asparagus, previously regarded as nothing more than an edible object, is a wondrous realm of the most tender, magnificent colours, with beauty and charm to rival the most fragrant flower, the most beautiful woman. He discovered that alongside and within the known entity there lay a completely different entity, accessible only to the eye, and this is why he sought to paint it. In short, it is not the 'what' and the 'how' we are concerned with, but rather the 'what' and the 'what' – the object we conceive and the image we perceive. People rarely observe the latter, because their desire, their interest drives them further, and that's why this beauty, this richness eluded them, and that's why they fail to recognise their asparagus (the object) in Manet's asparagus (the perceptual image). Manet saw only the asparagus and the air above it and the shadow, others saw only the edible asparagus without colour, without shadow, without air, because these are things you cannot eat. And only slowly did we grasp this misunderstanding and realise that in fact we are rich beyond measure, because alongside the world of objects that we know lies a second world, the world of the visible. And so some saw this painting as a revelation, the beginning of a new and richer life, while for others it was an affront, an absurdity.

It is this visible element that people are wont to dismember, disassemble, deconstruct, removing parts, observing this one fleetingly, that one more closely and a third to the tiniest detail as their practical interests compel. And so this visible beauty disappears from their view, eludes their awareness, and when they stand before a work of art they inevitably ask: what is it supposed to be? They don't understand that as well as objects there can be 'visible things'. It may amuse us to hear the maid who has just returned from the theatre with glowing cheeks talking about the wickedness or bravery of a particular actor. Not content with the surface of the art form, she tries to penetrate it, tries to reach the figure through the performer, who presents nothing more than a fine apparition of the figure, and so she destroys the work of art. People may mock such naivety, but they commit the same folly when they look at paintings, by concentrating so intently on the object, wanting to know precisely where the bridge ends, whether that bright spot in the water is a bird or a scrap of paper. They, too, attempt to penetrate beneath the surface, behind the apparition to the object that they know from their lives. They, too, believe, in all their naivety, that only a practical life has worth and the right to exist.

They seek the spiritual, the ideal, which can only be summoned conceptually, little suspecting that as an emotional experience the visible is of course just as spiritual, just as ideal, just as valuable as any other great stirring of the soul. Here, too, the most essential thing is not the object but the sensation, the stirring that it evokes. All that is good in our lives, all of our happiness is formed of the great joys, the emotions, that accompany the pinnacles of our actions, our existence. Everything that is, everything that happens we evaluate through feeling. So naturally those who have only experienced such happiness in the practical, objective world might believe a work of art to derive its value solely from its relationship to, its reminder of this world. But this ignores the fact that the world of the visible can induce immediate feelings just like the world of practical experience, and that this too has immediate value that confers immediate joy, happiness, life-enhancement, ideals.

Admittedly, only those who can truly see will reach this point; those who devote themselves utterly and unconditionally to the visible, just as they devote themselves to their work, their desires, their lives. Only those who yield to a desire with their entire soul, their entire temperament, can achieve great things, and so the wondrous world of the visible only reveals itself to those with the passion to seize it, those who have learnt to divine it, who have the experience and persistence to pursue it. Because a capacity for enjoyment is no more of a given than a capacity for work. It has taken humans an eternity to see that which has been passed down to us in the present day as a self-evident possession.

It is perhaps the most astonishing miracle in the formation of our lives that what appears as misfortune, unfavourable circumstances keeping us from our original objective, leading us astray from our desired objective – this is the very thing that marks the beginning of a new, unforeseen experience. So often we set out like Saul searching for his father's donkeys, only to discover kingdoms[10]. When people first began to depict, they aimed merely to reproduce the people and the things that were dear to them in their willed life, to remember them. And in their primitive way they initially drew only what was objectively essential, but soon discovered that there was much more to see in the depiction than what they had hitherto known only as an object, and there was joy in this revelation, beginning with the joy in that which lies closest to the objective – the form. They depicted the plastic form, which can be seen and touched, but largely without the contingencies of position, which have nothing to say in representation – stiffly posed, strictly symmetrical, but with all the attributes of clothing or insignia judged essential. What a long journey it is from the diffidently proffered left foot of the Apollo of Tenea[11] to the figures of Michelangelo, to the delight in overlapping, foreshortening, in unusual positions and perspectives that count for nothing in the realm of necessity. And then what a leap from the sculptural perception of Florence to the colourful reverie of the Venetians, to the delight in shadows, in contour edges, in refracting colours, in spatial masses. And then the long journey from Venice via Velásquez and Goya, through to Frans Hals and Rembrandt, to the modern French painters, to Manet and Monet[12], Cézanne, Degas and van Gogh. And it is only here that the visible becomes completely detached from the representational. The object itself is cast aside entirely, and if those biased toward practical reality in art were still able to evince interest if not genuine artistic sensitivity up to this point, because the subject, the actual *appeared* in the painting, the modern French painters ensure this is no longer possible. They consider only what is visually essential and occlude the object where it conflicts with this vision. For the first time we are seeing absolute, unconditional art. No more hiding, no more compromise. And this is why these things inspire such bitter, savage struggle.

It was the French who discovered the veil of air, which turns things into completely different entities with new laws and new forms of beauty. No longer were they painting people, bridges or towers, but the curious phenomena that air, light, dust and glare make of them. In the realm of activity we are obliged to see through this veil, to identify. However, these painters discovered that if you simply look, without destroying the appearance of the visible through either a sharp focus or deliberate abstraction, a wondrous new world opens up. And they painted what they saw. This explains the sketch-like nature of their paintings so bemoaned by pedants. But our objective for now is simply to learn, and not to criticise. It would be foolish to bring in 'patriotic' considerations here. We have learned from foreigners since time immemorial, inherited the foundations of our culture from antiquity, the Gothic from France, Baroque art and music from Italy, Shakespeare and industrial development from England. There is no shame in learning, but there is in the infantile rejection of a new cultural bequest. And if there are certain people today who endlessly, solemnly proclaim the German people's rejection of such foreign antics, the counter-argument is that the German people would be the most stupid people of all if they were to shut themselves off from new realms of beauty. Our country's climate, in particular, requires such ways of seeing. We would be foregoing immense riches were we not to learn from the French how to see these veils of air. Nonetheless our vision remains German; our cities, our landscape and our climate are so different by nature that this way of seeing will lead us to discover completely different things than the French have discovered in their country. Engaging with the French pictures has certainly opened up new vistas for me, and transformed Berlin, the city where I live, into a visual wonder that offers something different every day. I will try to give a sense of this, as far as words may reproduce such a thing, and so induce others to similar means of seeing and enjoying.

THE SCENIC BEAUTY
OF THE CITY

And so I deliberately leave aside all that is readily accepted, 'natural' beauty, of which the scorned city also has its fair share – park, hill, river and lake. Nor will I touch on the old architecture, the pretty old houses, the historic churches and the charming squares from days gone by, although they merit far more attention than they get. So few realise that even beggarly Berlin has a wealth of old architecture and civic works, that its old houses and churches would – if only we could push them together – constitute a fine old town of estimable proportions. Here I wish only to speak of the modern city, which – with vanishingly few exceptions – is hideous in design. The buildings are garish and yet lifeless, the streets and squares merely satisfy practical requirements without assuming a spatial life of their own; devoid of diversity or variation, they stretch away monotonously. You can walk for hours through the new parts of Berlin and yet still feel like you haven't gone anywhere. Despite the sheer desire for every building to stand out, to distinguish itself from its neighbours, it all looks the same. But even here, in this hideous heap of stone, dwells beauty. Even here there is nature, and landscape. The changes in the weather – the sun, the rain, the fog – all lend this woeful ugliness a strange beauty.

THE VEILS OF DAY

It is perhaps the fog that does this most insistently, and its beauty has always attracted at least some notice. It transforms a street utterly. It draws a gentle veil over buildings – grey, when clouds cover the sun on high, or warm, golden and bright when a clear sky stretches above. It changes the colours of the buildings, makes them milder and more cohesive; it smears stark shadows or removes them altogether, and these buildings, almost all of them groaning under absurdly overblown plasterwork, seem finer, flatter, more subdued. Even the Dom[13] – that frightful product of aimless handicraft run amok – strikes a wonderful figure on hazy autumn mornings around ten o'clock when the fog is visible and warm; the pointless recesses, the thousands of dissections and divisions disappear, infused with the fog that fills and swells the riven forms. Fog refines poor architecture, filling streets that would otherwise run on into infinity and transforming a void into an enclosed space.

What fog makes tangible and clear, with an

effect perceptible to even the unobservant eye, air does so more subtly, more softly, more inconspicuously, casting a hazy, gauzy veil over everything, almost continuously. Its density changes, and so this veil changes every day as well, sometimes almost imperceptibly and sometimes with very obvious effect. How beautiful it is when the whole street appears to be composed of a thousand gradations of grey and black, with the colourful highlight of an advertising pillar or the yellowing of an autumnal tree. How beautiful, when after a long dry spell everything seems entirely light grey, almost white. How wonderful it is on a bright summer day when the soft haze, visible only in the shadows, casts its fine, bright veil. Of course, as with nature, not everything is beautiful. You have to look for it. But this is harder, because unlike the open landscape, it hasn't been sought out and painted and described by thousands before us. Often it is only small sections that are beautiful, such as the reflective tram tracks in the grey asphalt or the recess of a loggia, whose red wall, half sunlit, half shaded, offers a delightful play of colour that contrasts with the grey of the facade. But expansive images can often delight as well – a happy accident of light, a fine distribution of shadow that stretches right across the street and transforms monotonous regularity into a dynamic form.

The effect of rain is completely different; it doesn't smudge colours but instead makes them heavier, darker, richer. The light grey asphalt turns deep brown, outlines become harder, the air clearer, the depths seem deeper, everything acquires firmness and solidity; but over this lies the wonder of lustre and reflection that envelops everything in a glittering net, turning the sensible, workaday street into a shimmering fairy tale, a sparkling dream.

Wilder and more fantastical still is the dusk; it thickens the haze of day, deposits steadily darkening clouds in the depths of buildings and appears to fill out the streets right and left; forms become calmer and heavier, every colour milder and more matt, everything gradually darkening, leaving only a few points of illumination – the colours of a wagon or a poster on an advertising pillar, so shrill and loud by day, now resounding bright and subtle in the descending grey. But the light of the sky drowns out everything, it dazzles the eye and spreads a mantle of flickering, uncertain, twitching light over the whole street, appearing everywhere and yet originating from nowhere. And then suddenly the sunset flares, everything that previously appeared grey and moribund now warm and glowing. The air itself is filled with warm, bright colours, every shade turns lively, the peaks of the buildings and churches glimmer in garish reddish yellow, and the radiant blue of evening spreads through the dusky streets. It penetrates everywhere, stronger than any artificial light, filling the narrowest streets, indeed this is perhaps where it is strongest. There is nothing like sitting in a city café at this time of day, looking down from the first floor onto the masses of people becoming ever darker, sensing the little patch of sky above suddenly flaring up and watching the blue tide fill the whole street, penetrating through large windows into smoke-filled rooms and for a few fleeting moments displacing everything – the newspapers, the menus, the conversation and all the petty concerns of banal existence.

Fog, haze, sun, rain and dusk – these are the forces that swathe our great rock eyries in splendid, ever-changing colours, melding their forms, making them more self-enclosed, even monumental; turning the most pitiful courtyards, the bleakest neighbourhoods into worlds of wondrous colour. They transform seemingly uniform piles of stone into a living and ever-changing entity. One individual could never exhaust its riches; it would be more than enough to experience what is offered by your immediate environment, your courtyard, your building, the streets you walk down every day.

From my study[14] I can see a high gable wall; in fact from my desk it is all I can see, the sky only visible when I go right up to the window and crane my neck. The wall is bare, made with inferior bricks, ranging from yellow to reddish, with grey, irregular grouting. But this wall is a living thing, a creature that changes with the weather – grey, monotonous and heavy on overcast days, lively and dynamic when it's sunny. The red bricks glow brighter than usual then, and the wall acquires a shimmering grain as the unevenness of the masonry becomes more apparent. Sometimes the sun shines on the upper part of the wall. Then it turns fiery and luminescent, and the lower part takes on a soft, subtle, bluish tone. Against the wall – I live on the second floor – stretch the tips of the trees in the 'garden' with their thin, shining branches. In summer they bear huge leaves – trees wish to live and the youngest leaves absorb energy from the sky early on – their heavy green is rich and full against the matt tones of the wall. But in autumn, when the leaves start to yellow, they shine with the light of the sun against the shaded wall, with a mild glow that renders the shadows cool and bluish. And as other leaves turn red it results in a wonderful image of tenderness; the glowing red of the leaves against the softer red of the stone. But when you look out onto the garden in the late afternoon, when a light mist envelops the trees, you can imagine yourself to be in a wonderland – the bright glowing leaves, swaying slightly in the darkening space before the violet shimmering wall, and the bluish dusk surging all about, alternately concealing and revealing. Then comes winter, the leaves fall, and one day the tip of the highest tree, the only part that sees the sun, rises ghostly and unfathomable like a golden whorl set against the wall shimmering in tones of red and blue.

And just as this wall reflects the life of the year for me, so too does the street on which my building stands. Every morning I go downstairs for a few minutes to observe its changes. Its length shifts constantly with the transparency of the air, the ends almost always enclosed by haze, and the buildings appearing taller or smaller, advancing or receding as sun and shade dictate. The grey of the footpath and the road, the green clouds of the two rows of trees and the black columns of their trunks look different every day, not always beautiful, but often so delightful that I can't tear myself away. And it's the same everywhere.

Nearby is a Romanesque church[15]. Frightful, simply frightful as architecture, confused in construction, senseless in proportion, absurd in detail, laboriously amassed from a thousand old treasures. In architectonic terms I couldn't imagine a worst spectacle. It is impossible to get used to it. And yet I look at its towers every day. Because every day the air and the haze transform them into new wonders. In rain and storm the stone roofs of the towers grow darker than the walls and gable, they dominate the streets all around, and I see them several times a day in the changing light. They recede, light grey in a grey sky, they advance, dark and menacing; after the rain they appear green, from certain sides even violet, and then they glow almost white against the blue sky once again. They look different from a distance, different from close up, different in the light, different in the shadow, different every hour and every day. They too are just one part of the living entity that constantly surrounds us in mysterious ways, for which we can only find meagre words like weather, or climate.

While you can witness transformation in the things you see every day, seldom trod streets and districts can imprint themselves individually through their charm or their scale. One of the most stupendous that comes to mind is an iron bridge along the Stettin railway[16]. Behind the station stretches a street that accompanies the railway track, to the right a row of five-story buildings without balconies – flat, charmless, formless. But in the distance rises a dark behemoth. That's where the railway curves a little to the right and passes over the street on a bridge 70 metres long. Here the street droops below it, so it appears that the bridge is almost touching the ground, the heavy, giant supports shifting against one another and forming a dark, vaulting mass which passes right by the last building and seems to race toward it with a roar. A mountain in motion, black and towering, it appears like a trumpet blast; your heart stands still when you behold the monstrous forces, the passion, the size of the hulking mass. There is only one thing to which I can compare it. It was in the harbour at Kiel[17]. The armoured cruisers lay at dock at wide intervals. And one of the ships had all its signal flags out to dry; it was the same passionate, horrendous roar, made madder still by the bright colours which ended in a piercing red, the whole a giant, blood-red coxcomb from the deck to the tip of the mast, swaying sluggishly, in monstrous contrast to the giant forms of the ships in their mute grey. On a similar scale but more disjointed is the huge arc of Gleisdreieck[18], part of the above-ground railway, which presents an unusual contrast to the slender, abstruse forms of the iron construction.

Then there is something completely different, glittering, almost playful, and yet overpowering – the hall of the Schlesischer Station[19], the colossal roof measuring 207 by 54 metres, supported by countless slender iron columns, so slender, that you can hardly make out where they join up, and they appear almost painfully sharp to the eye. As architecture it is hideous, but when fine haze fills the wide hall and turns the iron rods into an endless, glittering spider's web – the effect is unparalleled.

In high summer, the sight of certain streets in the north-east presents an unusual contrast. The buildings are very tall, taller than currently permitted, but without alcoves, hideously plastered with a thousand forms, devoid of understanding or vitality. There are two tall, gloomy walls, with a senseless profusion of cornices and mouldings spreading a network of black shadows where the sun strikes the surface, rendering the cloudy grey of the paintwork even heavier on the shaded side. But each of these buildings has two latticework balconies like little birdcages in each storey, each completely filled with the dark green and red of the flowers and creepers so diligently grown there. The street-side walls appear to be entirely decked with dense, richly coloured nests, which in the displacement of perspective perch right on top of one another, lending the cheerless, meagre street a strange charm, a glow of restrained passion, of fantastic magnificence. And so out of a dry paragraph in the building code, out of the most heedless exploitation of the available land, out of architectonic folly and the penned-in city-dweller's longing for flowers and greenery, arises an image of rare beauty. Naturally this is an unusually fortunate confluence. Grand impressions are more easily won where the gigantic proportions of engineering structures offer a certain monumentality even in their raw forms, especially in the great factory halls, although known only to a few, and above all in railway halls. How wonderful Friedrichstrasse Station[20] is when you stand on the outside platform suspended over the Spree, when you see nothing of the 'architecture', only the huge expanse of the glass apron[21], and its counterpoint to the petty jumble of buildings all around. It is particularly fine when twilight shadows fuse the tattered confusion of the surrounding area into one form, and the numerous little panes begin reflecting the sunset, the entire surface glowing and shimmering with life, arching over the low, dark, nocturnal fissure from which the broad, menacing form of the locomotive thrusts forth. And what a heightening of sensation when you enter the darkening halls yet filled with a hesitant daylight; the giant, gently curving form indistinct in the murky haze, a sea of grey, hushed shades, from the bright rising steam to the heavy dark roof cladding and the full black of the locomotive roaring in from the east. Above it, however, the evening sun picks out a gable at random, engulfing it in bright flames that shine in the dull surface of the glass apron like a towering, red, shimmering mountain.

THE VEILS OF NIGHT

And if the daytime offers a thousand colourful veils, this is especially true of the night. While starry skies and moonlight rarely make much impact, artificial lights bring an endless play of colour. Dusk finds them already blending in. What a charming sight – the bluish shimmering street under the expiring pink sky, the fine chiaroscuro that reduces colours to mute tones, when the long rows of greenish incandescent gaslights appear; at first barely visible, then like colourful spots and only coming into their own in the descending dark. Slowly the night pours into the street as though filling a vessel, right up the street-side walls, heaviest at the foot of the buildings. The glare of the deep blue sky increases the shady veils, and in this sea formed by layers of haze and shadows the colourful light begins its eternal game. There is a great variety of colours and luminosity. The green and yellowy-white of the incandescent lamp, the mild blue of the regular arc lamp, the red and orange shades of the Bremer[22] light and the new arc light, the red and white of the light bulb and the new metal filament lamps. Then there's the dark red and green of the vehicle lights. Every street offers new arrangements and contrasts.

A wide street like Hardenbergstrasse[23] can be wonderfully peaceful and expansive, with just two rows of bluish arc lamps, the whole slightly fractured street in clear, full light, uninterrupted by the shriek of illuminated advertising. To the right and left the buildings seem to recede in the half-light, and in the front gardens the trees appear stranger than they ever do in daylight, like mountains of moss, almost, their light-green tips standing out against the deep black background. The dark green clouds lurk eerily in the depths of the gardens, but where the trees meet the street and their thick branches extend over the footpath, the jagged forms of their leaves flare up, passers-by appear edged in glowing contours in the light that seeps through, all of it combining to form a glowing lace veil, a transport of delight in its delicate definitions, its rich density and its dynamism. And on the ground the fantastical network of leaf shadows contrasts with the cool shimmering stone in fine, warm tones. On rainy days, however, the image changes completely. The street turns dark, the smooth light grey of the asphalt becomes brownish like milk coffee, the glittering waves of its surface reflecting the lantern light. A fine fresh mist fills the air, and the entire sky appears to be covered with a wondrous veil of bluish violet.

The light is different in narrower streets where the rows of buildings are closer together, making the darkness palpable, the rows of trees enveloping the upper storeys in shimmering shadows, which to the dazzled eye appear shrouded in tender light. Bare and bright and bar all reflection lies the dry asphalt – it is only the tram tracks that glitter. But under the trees, where the lamps hanging in the middle of the street don't reach, from the lower storeys of the buildings, from the long rows of shops bursts forth a dense mêlée of bright light that renders people as black shadows. The buildings seem to sway in the air, and beneath them the gleaming flood of light pours out of what look like gaping mouths.

  A quiet side street, on the other hand, presents an impression of darkness. Where before a course of light seemed to run alongside the buildings, here the street seems entirely filled with darkness, and the few gas lanterns shine as if in little cages, which at the same time hollow out the air. They have an unsteady halo of light about them, but it barely reaches a few metres; they seem like mere points of light, not like the strong light sources that carve out giant vaults in the air, completely filling them with light. And when we step into these light vaults, the light plays all about, we are in a space delineated by a transparent yet clearly perceptible wall. There is a particular charm when you stand within one of these light spaces and see another further on, as though through a veil. Once in Dresden I experienced this very strongly, in the Schloßstrasse. Numerous red burning arc lamps fill the entire narrow street, forming a vault reaching to the third storey and on to the Altmarkt. But whereas before there was a shimmering bluish light, here the visual impression is muted by the reddish walls that surround you like soft music. Naturally it depends on the atmosphere. On muggy, dusty nights the cavities are smaller, after rain and wind they often grow unexpectedly enormous, indeed they almost seem to disappear. And it is particularly fine when weak lights stake out greater importance by casting their cones of illumination up high walls and creating great fields of colour. That's what you see at the aforementioned Romanesque church, where the adjoining streets all have electric lamps – only the square around is lit by gas. And now the bright limestone shimmers in a soft, murky green, and the whole church is cloaked in a dark mantle, isolating it from the busy streets, while the towers disappear from view into the low-slung night.

And then you can find still other effects on the canal[24], which is only dimly lit, and bordered by two quayside streets, each planted with three rows of trees. The thick treetops prevent light from making its full impact. The quiet buildings rise darkly behind the shadowy clouds of the treetops. The gas lanterns seem like points of light attracting cabs and automobiles; a fine, blinking web of stars spreads out above this dark mass. The smooth, turgid water is completely dark, and this silent, spectral mirror below reflects the gentle life of the night above, shimmering at the passer-by. And then this passer-by moves on and a bend suddenly reveals the splendid trumpet fanfare of the brightly illuminated Potsdamer Bridge[25], teeming with life.

THE STREET AS LIVING ENTITY

Where veils of air, dusk and artificial light turn the fatuous, desolate streets into fantastical formations their builders never conceived, with shadow and shimmer changing the sober, cuboid and linear into lively, lithe and generous forms, people and vehicles transform those mute forms into a living entity, one that may awaken, become active or weary, differing from day-to-day to holiday.

Generally we don't regard people as nature – quite the contrary, in fact. The modern moralist – who no longer preaches in the church, as a rule – is only too ready to assume that sin originates in youth and continues from there, the source of all that is unnatural and abominable. And so the philistine, lacking the knowledge required to understand the sufferings or transgressions of others – and lacking the inner joy to take mercy on others – anxiously avoids the rabble and calls down his curse from the safe distance of his hearth. And yet it is precisely in the city that you discover a side of people that is endlessly appealing, and that in the smaller communities must remain hidden. There everyone knows everyone else, and this 'everyone' is covetous and demanding. You have to converse on meeting, establish some kind of rapport. In the city you can stroll past hundreds every day, thousands, silent and distant, as you might walk past the trees in the forest. People are only apparitions, organisations in themselves, with inner cohesion that doesn't affect us, but with forms that are as accessible to us as the forms of mountains and trees. Humans are part of nature. And this part of nature is as charming, and as attractive, as any. What a wealth of types, gradations in age, in development, in the design of the body and the mind. It is only the foolish and the ignorant who regard the outer and the inner as divisible, but those who can see find that gait and posture, eye and mouth speak of an entire inner life. Not the tiresome sequence of external events that so attracts the attention of the curious, but rather life, entire, as a whole, its curious speed, its warmth, its tension, its complexity, its refinement, its momentum and its strength united as one, directly accessible to emotion. There are few things better than sitting on the tram in silence and watching people – not eavesdropping, but just experiencing and enjoying, observantly, sensitively. You can find so much beauty there, often quite still, inconspicuous, concealed from the unobservant by age, illness, mourning, severe pain, often splendid enough to conquer the blindest among us. Some peculiar people claim that a feeling for beauty stems entirely from sensuality. But this turns the matter on its head. Sensuality may rouse the eye, but the more subtly we can see, the more we delight in forms that would never inspire a sensual response. And so it is – much as the doctor may fail to understand – that the artist can depict the sick precisely *because* of their illness. How subtle the sickly colours of city children, what wonderful, austere beauty want and privation often bring to their features. And even depravity and impudence can have beauty, power and indeed greatness. The naive only see beauty where they feel desire. But those who can see may also find it where desire is mute. That's why they linger, enjoying and sharing unfeigned interest in those places from which the 'healthy', the 'unspoilt' flee in horror while loudly venting their indignation. The world would indeed be unbearable without the beauty of the weak, the old and the sick, and those who can find it may walk through the poorest districts without fear.

Admittedly it is far more amusing, and easier, to stroll through the streets of the well-to-do districts and observe the bright bustle of women. Female fashions, so often criticised, are almost the only form of design today that is lively and dynamic. To dismiss fashion as meaningless folly precisely because of its transience, as the pedant does, is a transgression against life. Because fashion is merely a symbol of life itself, constantly passing, changing, profligate with its gifts, not anxiously calculating whether this cost stands in reasonable proportion to the benefit. Nature spreads a thousand seeds all about, and perhaps only one of them will grow, and it is precisely this profligacy of ideas, this eternal beginning, this wealth of colour, that makes fashion so pleasurable. Doctors rightly denounce the lacing of the body, and anyone familiar with naked beauty will agree. But this will be of little value until sunbathing and mixed swimming reveal the beauty of nudity, and awaken the desire for it. Until then the reformers will have to seriously improve their efforts in taste and sensibility if they wish to counter common fashion. It continues to lead the way in its sense of colour, elegance, charm and self-assurance. Here we find the only thing that we have taken from visual culture in recent years – the sense of colour – asserting itself to positive effect. And instead of churlishly condemning its shortcomings, we should acknowledge how much more attractive the fabrics have become, how much finer their shading, how much more advanced the ability to match and layer colours to highlight a few solitary points. If lace and embroidery, detail in general, leave much to be desired, the whole is often charming enough, in any case more successful than most of the modern rooms that are praised as achievements today, but which have meaning only in their colour, and in which a poverty, indeed crudeness of form is naturally more apparent and more embarrassing than an ensemble that lends grace to the wearer's movements while rendering details blurry and forgettable.

But even if we refuse to accept this beauty, there is still the beauty created by people lingering on the street, regardless of the individual. It only takes one person, a moving point, to alter the impression of the orderly symmetrical street; it acquires a human, asymmetrical axis, to a certain extent, the open space is divided by the moving form, distance and scale assume new meaning. In the act of a person rearing out of the flat expansive plane of the street, this point assumes a particular emphasis in the perspective image, somehow becoming clearer in its spatial position; and because there is a certain average height known to all, we can perceive the space directly. The flat visual image, comprising nothing more than gentle shifts in depth[26], expands beyond. In their forms, humans create what the architect and the painter call *space*, which is something completely distinct from mathematical or even epistemological space. This pictorial, architectonic space is music, and rhythm, because it checks our expansion within a certain proportion, because it alternately releases and encloses us. Today's street makes for a miserable product as an architectonic space. It is improved by air and light, but a person walking divides it anew, animates it, broadens it, fills the dead street with the music and the shifting rhythm of spatial life. But there's even more. Because humans walk along the same street in different ways, when workers hurry to their offices in the morning, when women are out shopping, in the mid-morning, in the evening – streets break down into quiet and loud, into those for striding, strolling and looking around. The streets assume the life of their hours, they have good and bad sides; there are Sunday streets and everyday streets, each distinguished by the density, tempo and type of its throng. What seems grey and urgent one day is colourful and congenial the next.

And then there's the surge of carts and horses. Here, too, there are wondrous individual forms – a trotter, an English racehorse and the heavy workhorse with its thick stockings. Admittedly the vehicles only seldom achieve that sharp, lithe beauty we so admire in modern sailing boats, which require flawless lines, flawless materials and flawless construction. The cabs are conventional and dull, the automobiles still uncertain in their forms, the commercial coaches often curiously colourful and bizarre. You can't view them in isolation as objective forms, but they are appealing and handsome within the image, where foreshortening and displacement give rise to strange new formations, where the garish paintwork becomes softer in the veils cast over everything. This telescoping and clustering of forms is particularly noticeable in the dusk, when the shadow clouds of evening fill out forms. Horse and carriage become one, to the alert eye they appear as a grey mass with dark shadows and shining highlights here and there. Perspective seems to disappear completely, there is no front and back any more, the whole appears to be a shifting nocturnal mountain, with the red, dull, ghostly lights of the lanterns above. And so all of these vehicles become one wondrous living entity – the giant, yellow crate of the post coach, the wavering, thundering edifice of the automobile omnibus and the glass ship of the tram, their shining green bodies appear to glide away before suddenly turning a corner, the large pulleys casting off sparks.

They all help form the space of the street and contribute to the life of its hours. They stretch the street up and down, fill the space between the footpaths, hassle and harry within the dense swarm of the great arterial roads, disappear, sink into quieter streets. But wherever they go they bring movement, vitality. Even where they stand and wait, they give the street a new and unexpected appearance, constantly changing spatial perception by small degrees. This often results in truly handsome images. I have a particularly vivid memory of such a sight. It was during a hot summer on the Ringbahn[27] somewhere in the north, where the railway tracks on the bridges are not bedded down in muffling sand in consideration of the local residents and their ears, but instead lay hard and clattering on the structures. Beneath one such bridge stood a wagon with wooden beams, two heavy horses in front, their weary heads drooping. They stood completely to one side of the street before a yellowish brick wall and their position made the underpass opening higher and wider. On the other side were two children, who made the space even more tangible. Outside the scorching sun shone through a stuffy haze, and the brightness seemed to close off the space with a transparent mantle fore and aft, this space now filled with bluish shadows. But in the cool shade, thousands of individual sunbeams trickled through the gaps in the iron construction over the dusty street, as through the branches of a tree, over the children, over the yellow wood and over the silent, giant horses.

It is the life of the space that here, as in so many other cases, gives such a strong, meaningful foundation to form and colour, and it is difficult to provide a clear sense of it. If you think of architecture, you first of all think of the structural elements – facades, columns, ornamentation – but all of that is of merely secondary importance. The greatest impact comes not from the form, rather from its inversion, the space, the void that spreads rhythmically between its walls, is limited by them, while retaining a vigour that surpasses them in significance. If you can perceive space, its directions and its proportions, these movements of the void come to you as music and you gain access to an almost unknown world, the world of the architect and the artist. Because just as the architect is delighted by the play of spatial movement within the walls that he has constructed, so the painter delights in the diverse, convoluted forms of the space that in the landscape lies between mountain and forest, and in the city arises between people and vehicles on the plane of the street.

Among the most astonishing things in this regard is the life of the square. Opposite the unfortunate Romanesque church is a café[28] with a terrace, where I have often sat for hours on summer evenings, never tiring of the bright play of people coming and going. As architecture the square is absurd, perhaps even worse as a traffic facility – it is as though someone decided to assemble the greatest possible number of dangerous crossings in one place – but as a plane with people distributed across it, it is unique. Here the streams of people from the neighbouring streets dissolve in every direction, and the whole square seems to be covered with individuals. Each is at a remove from the others. The space between them expands. In perspective displacement the distant forms appear ever smaller, and you get a clear sense of the broad elongation of the square. Each is distinct from the other, coming together in greater density and then moving off to create gaps, the division of the space constantly shifting. The passers-by push past one another, conceal one another, break off again, walk free and alone, each upright and emphasising, elucidating a section of the square, and so the space between them becomes a giant, tangible living entity, more remarkable still when the sun allots pedestrians their own accompanying shadows, or rain spreads a sparkling, uncertain reflection beneath their feet. And amid this curious spatial life unfolds the swarm of bright painted coaches, the colourful clothes, everything melded, masked and made beautiful by the veils of day and dusk.

Such things have rarely been painted; crowds of people in paintings almost always fuse into shapeless clumps. Each may have a little free space ahead, but there is no living space between them. The shades of air call for a more sensitive, insistent seeing than is ordinarily required. I recall only one painting, by Monet[29], that reproduced the peculiarities of these phenomena. At the bank of a river lies a long barge, from which a number of parallel running boards lead to the shore, traversed by workers carrying goods. The consecutiveness, the perspective displacement, the diminution of the figures and their loose separation from one another, all of it is wonderfully represented. I witnessed a similar sight during the construction of a large hall, which stood there with raw walls and covered with iron supports, the windows sealed with boards against the cold, so that inside a half-light spread over the floor, which was made up of rhythmically arranged iron beams. Slowly, laboriously, over the planks that formed the walkway came a column of workers, each with a heavy brown sack of concrete on his back, and this slow, striding chain gave the desolate high space a solemnity beyond comprehension that made me forget everything around me for a few moments, including my work – the reason I was there in the first place.

But please, I beg you, do not think of Meunier[30]. It was not the dignity of labour – or whatever the overblown phrase is – that made this image so great. Meunier, like so many others, infused the worker with greatness, made a hero of him, made him Greek, because he couldn't see his actual beauty. These workers weren't moving about with the tensed muscles of an actor mimicking strain and strength, but were instead lugging gently and cautiously, having learned from experience that there is a long time until the end of the working day and that even the strongest must conserve his energies if he is to last that long. And it was precisely this slow, peculiar motion, one that we don't see in the theatre, that had beauty and charm, filling the space with solemnity made even greater because it was seen and perceived in an entirely new way. I have dwelt on this point at such length because I wish to avoid misunderstanding, and because it is something I have never read about before.

It is these effects of space and motion, combined with the veils of air and the light of the city, that together form the inconceivably bright, eternally inexhaustible fairy tale. From the wealth of impressions recent years have offered me, I wish to select just a few scenes that may give some idea of the splendour of these things.

At Westend there is a parade ground[31] that stretches out, limited at its southern end by a deep incision of the Ringbahn, to the north by villas[32] and their gardens, to the east by the approach of tenements – structural 'development' has already reached here – and to the west by another railway cutting beyond which spreads the low seam of the Grunewald forest. On sunny Sunday afternoons the enormous field is covered with people, broad streams of pedestrians burst out of the bridges and adjacent streets in dense droves, but the wide field lures them somehow, and the crowd dissolves, disperses, and a bright coming and going ensues. Several football games are under way simultaneously. The large playing fields are marked out with pennants, animated by the bright blotches of the players, who seem tiny against the enormous plane. The huge numbers of spectators aren't enough to form fixed rows. All is loose, stretching free into infinitude. A prevailing sense of joy spreads over everyone present. The life and laments of the individual disappear, the life of the collectivity becomes clearly tangible, assumes visible form. It is a wonderful thing to walk among people without thinking but simply experiencing the crowd instead. But even this is not the strangest thing. Most remarkable of all – strong, mysterious, as inescapable as fate – is the ground that bears this crowd. It is bright green with large brown patches that stretch for long distances and shift in perspective, becoming smaller in the distance and making the wide expanse tangible. And this ground gently, expansively undulates and with it the tattered rug of people. They're playing, running or strolling, moving this way and that – and without knowing or suspecting it they collectively constitute one wonderful, enormous form, a form borrowed from the bare ground and yet stranger and more impressive; colourful, rich and dynamic, criss-crossing each other in their thousands, yet in their movements deferring unconditionally to a secret law, the law of the soil. Naturally I mean this without any of the literary symbols it may conjure. It is not that to which it might be compared but the thing itself that makes this image, the way our eyes capture it, so strong. And it proves that our eyes may directly experience that which we ordinarily imagine can only be captured in a notional, poetic way. And it retains its splendour as stratification and subtle order come to the confused mass in the twilight, when everyone heads home, crowding on to the bridges leading back to the city.

The bright mass is like a forest and just as invigorating and miraculous for those who remain still, who can see and succumb. Sometimes, on the hottest days, when the well-to-do flee Berlin, when stifling haze and hot motionless air make it almost impossible to stay indoors, I would go walking through the city of a Sunday, down Unter den Linden[33], to witness the great homeward journey of the polished people. There are almost no carriages on Frederick the Great's beautiful forum[34], and the bright giant asphalt surface lies there unused in Sunday langour, almost like the stone slabs of the Piazza San Marco. But people rarely tread there, as if in weekday reservation, and so only a thin network of people spreads across the wide plane, and the two colourful streams cluster at the sides, completely filling the evening shadows that gently appear at the foot of the buildings; black hats, pale suits and the bright flash of the women form a dense, shimmering band that glides across its bright, open parts like a colourful snake seeking the shelter of a building. The whole is submerged in a wondrous light; everything appears to be enveloped by reddish dust, the pale blue sky shimmers through, the buildings slowly grow pale, by now only struck by the sun here and there; they glow tenderly and insistently, the gentle dynamism of their structures delimiting and forming the wide space. Out of the dark grey in the distance a few windows in the palace[35] reflect the sun; the garish yellow of a post carriage sparkles below.

Potsdamer Platz[36] in the evening. The two large lighting towers with their glittering red arc lamps hollow out enormous peaked domes in the thick, heavy air. The small, squat side streets spill onto it; the eye seems to perceive their ends no great distance away. Especially low are Potsdamer Strasse[37] and Bellevuestrasse[38] where the trees delimit flat vaults below. Their foremost crowns shimmer in garish lights like green cliffs with a thousand recesses, while the enormous masses of linden trees on Leipziger Platz[39] form dark, mute, distant mountain ranges. The crests of the buildings shine in shades of red and violet. People surge on the footpaths, and on the bright, dry asphalt the trams and carriages throng in eternal recurrence[40]. Their hoods glisten in the light, but below they appear entirely draped in a darkness overlaid with a whitish mist of glare. Sometimes the surge grows so large that there is barely a free space, and pedestrians crossing the roadway seem to rear up from the sea, as if from waves composed of wheels and horses' legs. And then quite unexpectedly a vehicle stops right in front of the onlookers and as if by magic it appears large, and decidedly tangible, where just now it seemed just another confused and ghostly something lost in shades of grey and black. Suddenly it disappears again, a giant horse's head fills our field of vision, its nostrils flaring wide – the animal breathes deep with labouring flanks – lending the profile of its head the nobility of an ancient equestrian bronze. Then just as suddenly the jumble dissolves, the little piece of asphalt before me seems to stretch into infinity, the dark, confused corral replaced by a shining bright spot, immediately consumed by the dark tumult once again. In this sharp, unaccustomed light the vehicles assume fantastical shapes; a stubby automobile cab appears like a giant bumblebee, under the bulky box of the carriage the darkness makes the slender spokes of the carriage wheels appear even thinner, calling spider's legs to mind, while the lanterns seem to float freely above the black mass. And below them all stretches the ground, like a mad realm of shadows scurrying spectrally across the surface in tireless vivacity.

In front of the Brandenburg Gate[41] on an autumn evening. The sunset is extinguished, and after dusk the sky is completely filled with that mysteriously penetrating blue. The thinning trees form light, reddish masses. Here again two large light masts cast a garish red light, reflected from the high columns of the gate. Through the openings you can see Unter den Linden in the dark, cooled by bluish illumination, with the night drawing in above. Only a few cornice edges of the little gate houses shine strange and sharp against the bluish scene between the columns, struck by the red light flowing over the gate. The wide square is completely enveloped by the reddish veils formed by the glare. The carriages seeking entrance look as though they are moving about the ocean floor. All tangible reality is lost to them. They seem to be formed of clouds, shadows and light. And between them the endless throngs, crossing the square from every side, returning from the Tiergarten[42]. The crush of carriages and people merges completely. The whole square is densely packed and seems to form one unified entity. Not even the cumbersome tram terminal can break the spell, so powerful are the light and air that envelop everything, bind everything, blend everything into one restless moving behemoth. And despite the thousand cries, everything seems quiet. The light drowns out the noise, you don't notice it. But a sense of the monstrous comes from the marble balustrades and walls which are so hideous by day. They form the banks of this lake of light, with countless people in repose, sitting, standing, barely recognisable in the hazy light, a silent, marvelling, festive gathering. With this the wide circle assumes true majesty, a beauty, a grandeur, an exaltation the equal of that which is most noble, with a power over the soul possessed only by the greatest of that which has been passed down to us from the past.

These are just a few excerpts, rendered in sketches. It is a world that should be depicted by poets who can apply the entire force of their art to finding the most descriptive, plastic, vivid words for this miracle. It is a world that should be painted by artists who can most ably and directly lend it form and colour and space.

None of this has been painted yet. Our painters still – and here this means painters in the sense previously discussed – go to foreign countries in search of subjects. And it is understandable that students of art learn in places where their teachers painted. And if landscapes and people should appear more picturesque there, this is only human, because they have already been conquered, pictorially speaking. I'm not familiar with Holland and I've never seen silvery Paris, and so I can't compare either. But I know enough French paintings and have the feeling that different yet comparable beauties await discovery and conquest here. Naturally this objective will only be achieved through long study and experimentation, it will take generations of painters before we have a sense of the extent of this world. But only then will the beauty of the metropolis become a self-evident heritage, like the beauty of the mountains, the plains and the lakes, only then will children grow up in sure possession of this heritage, as we grew up in the sure possession of the beauty of the landscape. And only then, with this firm foundation of visual pleasure, can we hope to witness the rise of all-encompassing design.

NOTES

[1] Henning Berger (1872-1924) was a Swedish writer who spent much of the 1890s in Chicago, the setting for his novel *Ysaïl*, which first appeared in Swedish in 1905. Endell quotes here from the German translation which had just been published.

[2] The trope of Germans as a *Volk der Dichter und Denker* (people of poets and thinkers) had been circulating since the Romantic era.

[3] *Gemüt* in the original, usually translated as 'temperament'; since around the time of Goethe, *Gemüt* was often spoken of as a uniquely German characteristic.

[4] The term *Heimat* is sufficiently different from 'homeland' (the most common translation) that it is left in the original throughout; see Afterword.

[5] This is one of the few specific, non-Berlin locations mentioned in the book, and one which curiously appears twice in different contexts. Schloßstrasse was a narrow street in Dresden's historic centre running between the eponymous Schloss (palace) and the Altmarkt (old market place). Most of its buildings were destroyed in the war.

[6] Otherwise known as Litfaß columns, these round structures used for advertising were first introduced to Berlin by printer Ernst Litfaß in the mid-19th century, and were intended to stem the unregulated practice of affixing posters in public space.

[7] **Author's footnote:** In 1905 I published a series of short essays in *Die Neue Gesellschaft* at the instigation of the publisher Dr H. Braun, which can be viewed as a complement to the statements here: On Vision. Evening Colours. Spring Trees. Treetops. Potsdamer Platz in Berlin. The

Art of Impressions. Vol. 1 1905. Issues 4, 7, 8, 10, 12 and 23 [included in this edition].

⁸ Also known as the 'yellow spot', the macula lutea is, as Endell indicates, the part of the retina that records the sharpest impressions.

⁹ This painting of white asparagus, dated 1880, was in the possession of German Impressionist painter Max Liebermann at the time Endell was writing, and is now held by the Wallraf-Richartz Museum in Cologne. It was originally painted for French art historian and collector Charles Ephrussi, who overpaid for the work, whereupon Manet made up the difference by sending him another canvas – depicting a single spear of asparagus.

¹⁰ 'And the asses of Kish Saul's father were lost. And Kish said to Saul his son, Take now one of the servants with thee, and arise, go seek the asses. And he passed through mount Ephraim, and passed through the land of Shalisha, but they found them not: then they passed through the land of Shalim, and there they were not: and he passed through the land of the Benjamites, but they found them not.' (1 Samuel 9.3-4)

¹¹ Now known as the Kouros of Tenea, this was a 6th century BC Greek statue of a smiling youth which Endell most likely saw in the Glyptothek in Munich, where it had been exhibited since the mid-19th century.

¹² Two of the three artworks reproduced in the original 1908 German edition of *The Beauty of the Metropolis* were by Claude Monet (the other by Max Liebermann). As well as his study of coal carriers (referenced later in the text), Monet's 1867 oil painting *Saint-Germain-l'Auxerrois Paris* was included as the frontispiece.

¹³ The Dom (Cathedral) zu Berlin was built on the site of places of worship dating back to the city's medieval origins. The present Protestant cathedral was designed by Julius Carl Raschdorff and consecrated in 1905; shortly afterwards Endell wrote an article about the new building in the journal *Freistatt*.

¹⁴ At the time of writing, Endell was living on Fasanenstrasse in the Wilmersdorf district (the address book for 1908 further reveals that the architect already had a telephone connection). This was also the site of his 'Schule für Formkunst' (school for the art of forms/design school). The building appears to be largely intact, along with the wall that Endell describes, although it has since been rendered.

¹⁵ The Kaiser-Wilhelm-Gedächtnis-Kirche (Memorial Church) was built to plans by Franz Schwechten on the orders of Wilhelm II to honour his grandfather, Wilhelm I, and opened in 1895. Endell's poor opinion of the structure (which he only ever refers to as 'the Romanesque church') also finds expression in the *Neue Gesellschaft* article 'Evening Colours'. Largely destroyed by Allied bombing in 1943, its truncated central spire was retained as a memorial.

¹⁶ The Liesenbrücken (officially classed as dual bridges) were built

between 1890 and 1896. A 1927 painting by Gustav Wunderwald, *Brücke über Garten- und Ackerstrasse (Berlin N)*, provides a vivid impression of the structure's imposing bulk. The bridges abutted the Berlin Wall during the separation of the city and lie intact yet unused in the present day.

[17] At the time Kiel was a major centre for the German navy that Wilhelm II was building up in his quest to challenge Britain's supremacy over the high seas.

[18] Gleisdreieck, still in use today, is an above-ground junction for two lines of the largely underground U-Bahn system. At the time the impression of dynamism it presented was magnified by the vast expanse of adjacent railway lines and sidings connecting with nearby Anhalter and Potsdamer stations, since destroyed. The year that *The Beauty of the Metropolis* was published saw a serious accident at Gleisdreieck in which two trains collided, with one carriage falling from the elevated railway to the street below.

[19] The version of this station that Endell knew, in the Friedrichshain district, was designed by Adolf Lohse and dated to 1867. This building was destroyed in the war but a rebuilt structure later became East Berlin's main passenger station (Hauptbahnhof), and in its new post-reunification incarnation it is known as Ostbahnhof.

[20] Designed by Johannes Vollmer, this central Berlin station opened in 1882 and retains much of its original form.

[21] **Author's footnote:** A finely wrought expression used by engineers to denote the end wall that hangs free over the tracks of a station hall.

[22] This light, famed for the strength of its illumination, was designed by the eccentric inventor Hugo Bremer (1869-1947).

[23] This street leads from the Kaiser Wilhelm Memorial Church to what is now Ernst-Reuter-Platz. Halfway along it is Steinplatz, the square for which Endell designed a residential block in 1907.

[24] The Landwehr Canal at the southern edge of central Berlin; confusingly this part of the text was illustrated with an 1899 pen and ink study by German Impressionist Max Liebermann (1847-1935) that depicts a canal in the Dutch city of Leiden.

[25] A key crossing of the Landwehr Canal, the Potsdamer Bridge of August Endell's time dated from 1898. The traffic was notoriously heavy here and the road difficult to cross; a tavern on one side catered specifically to pedestrians who had abandoned the attempt. The bridged was destroyed in the war and later replaced with a more functional design.

[26] **Author's footnote:** The dimension of depth is only seen approximately, indeterminate and wavy, in contrast to the other two dimensions.

[27] This railway encircles central Berlin, originally dating from 1877 when it was built to connect the city's terminal stations.

[28] The Romanisches Haus was built in a similar 'Romanesque' style to the

adjacent Kaiser Wilhelm Memorial Church by the same architect, Franz Schwechten. The café Endell would have frequented was later replaced by the famous Romanisches Café, which was a key meeting point for the city's artistic and literary avant-garde during the Weimar Republic.

[29] Monet's *Les déchargeurs de charbon*, also known as *Les charbonniers* (The Coalmen; 1875) was reproduced in the original edition of *The Beauty of the Metropolis*. Architectural historian Matthias Schirren suggests that Endell may have translated the motif of the coal-carriers into a section of his tilework facade in the Hackesche Höfe.

[30] Constantin Meunier (1831-1905), a Belgian artist who often depicted labourers

[31] The site of the parade ground is now occupied by the ICC trade fair complex.

[32] Endell was involved with a number of projects in the 'villa colony' to the north of the parade ground, either new villas or renovations. Most of them are still standing albeit in modified form, and one has been reconstructed.

[33] Unter den Linden was and remains Berlin's most prestigious boulevard, leading from the Brandenburg Gate to the River Spree.

[34] The Forum Fridericianum was a large expanse straddling Unter den Linden dominated by major cultural institutions, including the Staatsoper. While the buildings largely remain, the open space has reduced to what is now Bebelplatz, between the opera and the old library of the Humboldt University.

[35] Berlin's Stadtschloss (city palace) was built in various stages between the mid-15th and mid-19th centuries, partially destroyed in the Second World War and then demolished in 1950, with a reconstruction due to be opened in 2019. In an open letter defending modern art and criticising the Kaiser's reactionary and controlling tastes (see Afterword), Endell noted that 'There is no pathway to the *Schloss* for our art'.

[36] As many people pointed out at the time, Potsdamer Platz was more an enormous intersection than a square. Almost entirely destroyed in the war and later clearing operations for the construction of the Berlin Wall, it was re-built after reunification.

[37] Endell would not recognise the initial stretch of Potsdamer Strasse today. In his time it was a straight line running between Potsdamer Platz and the Potsdamer Bridge (a fragment survives as 'Alte Potsdamer Strasse'), but after the Second World War its route was radically altered by the architect Hans Scharoun to accommodate his Staatsbibliothek (library) and now describes a curve separating that building from the Philharmonie (concert hall), also in his design.

[38] A short street connecting Potsdamer Platz and Kemperplatz at the edge of the Tiergarten

[39] Adjacent to Potsdamer Platz, the octagonal Leipziger Platz was one

of three public squares laid out in different formations when a customs barrier (the 'Akzisemauer') went up around Berlin in the first half of the 18th century, the other two being the rectangular Pariser Platz and the circular Rondel (now Mehringplatz).

[40] A reference to Nietzsche's concept of 'eternal recurrence'

[41] This version of central Berlin's ceremonial gate was erected in 1791, when it formed part of the customs barrier.

[42] Originally royal hunting grounds, the Tiergarten received something like its present landscaped form in the designs of Peter Joseph Lenné around 1840.

ARTICLES FOR
DIE NEUE GESELLSCHAFT

ON VISION

The objective of all the arts is beauty. And beauty is nothing but the immediate sense of strong, intoxicating joy produced in us by sounds, words, forms and colours. We must simply learn to surrender to them entirely, so completely that our minds hold nothing but these forms, these sounds. But this training is more difficult in some arts than others. Curiously it is most difficult in the area that seems most accessible – verse. Poetry appears to depict, to instruct, and when we discuss it we use terms like morality, world order, reality, truth, patriotism, and so forth – but rarely beauty. Hardly anyone knows what constitutes good verse. It takes long, arduous work, great knowledge of literature – but even more so of life – to break free of the tangle of misconceptions that currently prevails. Music, however, is the most accessible of the arts, because it is not linked to thoughts (or if so, only forced thoughts). It compels us to listen, and only to listen. But unfortunately external factors make it less accessible; we must hear plentifully, frequently, and that requires money, or a proficiency in piano that few achieve.

The visual arts, on the other hand – these are available directly, to all of us. Seeing is something that can be practised anywhere, at any time. There's more than enough material. Buildings on the street, museums and churches, and above all nature – they all offer endless opportunities for training the eye. And nature is a good place to start. Particularly as all of us have had our minds more or less poisoned by the enormous amount of rubbish that has been written about art. Anyone who can view nature with discernment and sensitivity will soon find that works of art present no challenge. But conversely, anyone unable to see nature will not be able to see art. This is where the conflict between viewer and artist usually arises. Standing before a painting is a man, perhaps a successful businessman. But this man has never opened his eyes. He knows that the sky is blue, leaves are green and clouds are white – why should he bother looking at them? Naturally he takes offence when the artist paints the clouds green, say, or the leaves blue and the sky red. The only way for them to reach agreement is for them to both go out and really look at the sky and the earth. And it soon becomes apparent that there are numerous strange things in trees, in cab horses, in dogs' ears, things hitherto unimagined.

But how do you begin seeing – truly, artistically?

The first step is to entirely avoid thinking and conceptualising. Our minds must contain nothing but what we are currently viewing – the face of an old man, for instance. Why does it appeal to us? Not because it's the head of an old man and old age is venerable. Perhaps this head isn't venerable at all; perhaps it is evil, with little glinting eyes in deep red sockets, but there is a beauty to this evil. And wherein lies this beauty? Once again, it is not that experience teaches us that a man with such eyes is evil, and that evil is beautiful. Because that certainly isn't always the case and only rarely does our experience help us here. We don't even need experience for this. When we look at those eyes, we instantly sense evil and the attraction that it holds without having to think about it. In animals and children we call this instinct, but it is something we retain as adults as well. Seeing and feeling occur simultaneously. It is these instinctive, spontaneous feelings that count. All of us have them, it's just that we rarely pay attention to them. Naturally there are things that just as instinctively repel or bore us. In fact that is usually the case. Because despite what the sentimental and sanctimonious would have us believe, nature is not always beautiful – on the contrary. But there are elements of beauty everywhere, you just have to learn where to find them.

So first of all clear your mind, keep to the matter at hand, look closely and slowly.

This is by no means simple. We tend to regard seeing as easy. If ten people stand before a house they are all seeing the same thing. That's because the house is reproduced evenly across the retina of the eye, and what is reproduced there is what we see. And yet it is still not that simple. If I am sitting in a concert hall with my eyes open, listening intently, or crossing the street while engaged in lively conversation, there is much that is reproduced on my retina, but I don't notice it. The mind cannot execute numerous procedures simultaneously, it cannot see and talk and listen at the same time, but rather one after the other. And that's not all. The retina may well reproduce the whole house. But not every part of the retina sees with equal clarity; the lateral parts are particularly poor and blurred. There is only a very small part in the middle that sees precisely and sharply – the 'yellow spot'. And if we really want to look at a house, we have to apply this yellow spot to every part in turn, moving our eyes up and down, to the right and left, more or less scanning it line by line as we describe vertical and horizontal strokes across the house. It follows that we achieve very little through mere looking, and that it takes a fairly long time. The connoisseur requires hours to really look at a single painting, in which time the artistically disinclined might stroll past a thousand. But nor is it enough to simply glide over the forms. You have to follow the shapes, consider how they relate to one another. So this may be the crossbar on the window, then the row of windows, and the cornice above them. Or the bridge of the nose, the onset of the cheek, then the forehead, the eyebrows, the eye sockets, the oscillation of the upper lids, the eyelashes, the line of the lower lid, the eyeball, the forehead, then the triangle formed by the nose and eyes as a whole. Then just as closely the mouth, then mouth, nose and eyes as a whole, then chin, cheeks and forehead, and so forth. Naturally it takes practice, and as simple as it may sound here, it takes a long time until you can do it, and until you notice how much pleasure it provides. Because that is all it comes down to. Of the great amount that the retina may potentially offer us, we usually only see the small amount that is of practical interest to us – the front door, the door handle, the steps, the balustrade, the name plate. Everything else remains indistinct, hazy. But if we wish to see in an artistic sense we are not looking for things that are of interest to our professions or other purposes, instead we look to see whether there might be anything among the many forms and colours to bring us pleasure and enjoyment. And of course that is a completely different way of seeing.

Later I will offer a direct, practical introduction using certain examples to explain how to see, and what there is to see here and there. All I will add for now is that the reward for this effort is enormous. There are so many beautiful things directly accessible to us, and yet so few of us see them and enjoy them. A world of wondrous delight, under our very noses, so exquisite, so colourful and varied that they render invented fairy tale worlds obsolete. Today, the present, this reality is the most fantastical and incredible thing of all. Any wonders we might invent are meagre and slapdash in comparison. And foolishly we leave this treasure untouched. We don't need another world above the clouds or in the past; our world and our time hold an immense realm, one that may be closed to the naked eye, but open and available to the eye that sees.

EVENING COLOURS

Sadly we have every right to complain about the ugliness of our cities. Aside from the old parts – very little of which remain in Berlin – there can be no doubt that our streets are cheerless and desolate and abhorrent. But when sentimental folk declare this to be nothing more than the just result of an overly refined culture that has eliminated nature, entirely and artificially, this is both distorted and stupid. Certainly the concentration of people in cities is lamentable from the perspective of public health. But nature can never be eliminated, not even from those wonderful giant eyries of the stone caves in which we dwell. And there is great appeal in following the strange natural life in these stone chasms with seeing eyes.

The architecture is, of course, appalling; we turn away from it, shielding our eyes. But there is a ceaseless play of air and light about these nasty facades and streetscapes, constantly altering their appearance. The shadows shift continuously, their colours changing with the sun, with the blue sky, with the veils of mist and the glare of white clouds. And anyone who pays attention to these things will find great beauty amid all the unspeakable ugliness. Let us begin in broad terms. It is evening, and finally we find time for observation. The sun is going down somewhere, but we can't see it; as it happens none of the streets we can see leads to the west. We sense the ebbing of the light. The colours are turning pale, losing depth. Gradually, at the base of the row of buildings, the streets fill with darkness that slowly creeps up over the road and along the facades. Only the sky is glaringly bright, sometimes even obscuring our view of oncoming pedestrians. The plasterwork of the buildings disappears as a soft dark cloud seems to spread over all forms. Only the roof lines are clearly offset from the sky. And then something remarkable happens. During the day there are so many frightful gables and turrets that aim for splendour and achieve only tedium. But now we see nothing of their sculptural forms, they all meld into a strange silhouette, like the outline of a pine forest. On the square there is a modern Romanesque church – an opulent yet hopelessly absurd imitation of wondrous structures of old. Even it becomes beautiful in the evening when its silhouette contrasts with the sky, wrinkled and curiously busy. Lines form and flicker dimly against the luminous air, lines that the architect never dreamt of or suspected. Now the lanterns flare up and a sharp yellow light penetrates the darkness down below. And with it begins a fairy tale brighter and more delightful than any we heard as children. The colours return, but brighter and softer than they are in the daytime. Some points are still illuminated by the sky, like the colourful ladies' hats, some only by lanterns. But the artificial light lends them a different hue. A wall appears in luminous yellow, the asphalt stretches out in shimmering grey, the gentle bulges in its surface brighten, polished here and there by wheels, between them flash the tracks of the tramway. But finest of all the bright colours are those of the Litfaß columns, the merchants' wagons and women's clothes. Far too shrill by day, they now turn matt and delicate as Berlin's uncommonly beautiful, gauzy air comes into its own. Everything blurs into everything else, and out of the tones of grey and yellowy-brown advances a blue, a scarf of yellow, a wide waving strip of green. All things hard and true fuse into a fairy-tale veil, airy and transparent. The eye is dazzled by the lights, no darkness around them, a gentle, light grey in numerous subtle gradations spangled with shimmering fine points of colour. And all of it is in motion. This hardly seems like reality at all. An illuminated tram passes through the scene. But it too flees like a shadow. It casts unusual reflections as it glides around the corner, its big broad windows that a moment ago offered a view straight through to the other side of the street suddenly reflect a harsh flood of light.

And above it all the dark, glowing sky. It has turned quite sombre. But – fresh miracle – its colour, scarcely blue any more, lays a wondrously deep and passionate glow before eyes just now flooded with artificial light. I wouldn't dare explain it. It is impossible to describe. It is always amazing how powerful darkness can be.

An hour later the sky is completely black, with

only the gentle twinkling of the stars, or the vivid glow of a planet penetrating the dark cover. The buildings are all lit harshly from below. Strange contours of light form on the cornices and corbels. Circles of light from the lanterns segment the street. From the rows of trees, young spring green softens the yellow light and weaves green veils. This green glares brighter as we advance toward a light; it takes on a strange luminous tone that creates a wondrous impression against the dark street. The tree trunks are entirely dark, yet even here the glare weaves an almost imperceptible veil. But then there is a tree lit directly by gaslight, its bark shining grey, almost silver, deep black reserved for the contours where the branches curve away from us. An image that holds the most tender appeal, which although entirely 'unnatural' is still a thing of nature, above all a thing of beauty.

SPRING TREES

The time of new green growth. So often extolled, so 'poetic' in the public imagination. But unfortunately our poets rarely have eyes; nature gives us such tremendous abundance, miraculously infinite, and what people actually see of this is so laughably meagre – nothing more than what others have seen and extolled before. Spring arrives suddenly one day, says the unseeing city dweller, entirely ignorant of the slow miracles of the bud and the young leaf, and for all that, he's proud that he, enlightened and rational man, can still perceive this miracle. But it is a meagre miracle – speed invisible to the eye, a truly urban miracle that somehow brings to mind electric trams and other fine things. And yet the greatest of the miracles of spring is precisely the long duration, the transformation that constantly surprises anew. It is not that a bare tree bears foliage from one day to the next that is astonishing. No, the wonder here is how the bare winter tree assumes a hundred forms, a thousand, even, before it becomes the full, mature tree of summer. Slowly, slowly swell the buds of brown, their colours cast off deathly winter as they lighten slightly and their skins tauten. And already the tree is a different tree, thick nodes revitalising the bare twigs all over. Now the tip of the bud breaks out, the dark bract slowly parting to reveal thin, bright membranes within. The nodes assume spherical forms with glowing points at the tips. And should we approach the tree from afar, it appears as though a barely visible shimmering web has been spun about the dark, naked twigs. Each day the buds swell a little more, their tips now a bright soft green, the web about the tree grows thicker and more lustrous. Now the young leaves break free of the bud, already complete yet still folded up, forming loose green spheres with rounded bracts. And yet again the tree becomes a different entity. The gentle tenderness is gone; for all their youth the spheres of leaves have something powerful about them, they are no mere embellishment to the twigs as the buds were, but rather form a counterweight. The spheres expand, their forms dissolve, the leaves diverge and form large unified patches that continue growing toward each other. Yet between them you can still see the blue or grey sky. We now have a completely new tree. In winter it was thin and bare, hard in its lines, dull in its colour. Now it is light and full, round in form and a sigh of green. A completely new creation that scarcely seems kin to the other. And it becomes fuller and lusher all the time, the colour deeper and stronger, the twigs disappearing as a large spherical green cloud floats atop the dark trunk.

But this is the merest outline. There are not enough words to describe all the transitions, to make them so vivid that you can see them with your mind's eye. I can only spark your curiosity, can only entice you to look at one tree, closely, day by day. For weeks it is a new entity every day, changing a hundred times, surprising the attentive observer with its constant twists.

Here we don't have just one genus, we have many, and each is different in character from the others, and even within the same genus each tree is different, each has its own personal features, its own idiosyncrasies accorded it and it alone. Watch closely and you may soon find friends and favourites among trees, just as you do among people. But then you have to see them individually, rather than trying to see them all. See slowly. See, and wait. Soon you will gravitate to one here or there, in wonder, and experience joys unknown. It is only the profusion that confounds, diverts the attention, fatigues the eyes – the mad exultant tumult of spring.

First let us look at the basic form of the trunk and branches. You can still make them out. We encounter different types all the time. But you shouldn't try and categorise them botanically. This would be quickly accomplished, and yet you would have seen nothing. There are trunks that rise steeply without bends, cast out strong individual branches and then all of a sudden a cloud of thin straight twigs. Just next to it is another, it too has a tall trunk and a few strong limbs, but these limbs curve outwards, and the twigs are hunched with hard, sharp corners, like grasping fingers. Still others are heavy and round in the trunk, suddenly casting out numerous branches in all directions, later diverging in myriad ways to form a thick sphere. The treetops assume all manner of forms, but they almost always recall clouds, heavy spherical forms, low cowls, angular and irregular, leaving the supporting branches completely free. Then there is the heavy curtain of the fir tree, the dark shuttlecock of the pine, the long flowing hair of the birch. You can see all of this in thousand-fold transformation right here, on our streets, in our ornamental gardens and public parks. And that is nothing but the most rudimentary introduction to seeing. When we learn to pay attention – that's where the finer pleasures begin. Before us stands a giant tree, its trunk veering a little to the left, its main branches extending back to the right in forceful oscillation, as though the wind had bent the twigs the other way and the tree had rendered its opposition. This gives rise to wonderful, powerful lines, bold, passionate and expansive, and when this giant takes the first green veil it provides a wonderful contrast, a blend of strength and tenderness. Next to it is a beech trunk, all in silvery grey, smooth, with unusual furrows; a few metres above the ground it suddenly veers to one side; a branch has been sawn off and now the bulky trunk is growing bolt upright, into the sky, the force of its growth redoubled by the severe deflection that precedes it.

And then the green! How meagre is language, having just one miserable word to denote this endless variety. It begins with a bright, almost white green, but another shimmers almost yellow, another bluish, still others are light brown and pink. Gradually the colours deepen, the surface of the leaf tautens. Some begin to shine, many remain dull, others are covered with white down, still others have fine tufts at the edge. Topside and underside differ in almost every case. The shade changes from day to day, from tree to tree. Even in the long rows of streets, where every tree is the same species and the same age, a few stand out, stunted by chance, and form charming contrasts to their comrades. What soft, fine beauty and what profusion, what pleasure. And then over this wonderland steals the dream of the blossom.

TREETOPS

I really must talk about spring trees once more. What seems lively in writing transpires as lifeless and meagre once you go outside, and so I wish to start again from the beginning, to capture this bright wonder, to induce the reader to observe and to enjoy. This abundance can never be exhausted, with each year and each day bringing sights never before seen, even to those who immerse themselves in watching these things for years. Yet it would be foolish to speak of the beauty of the sky or the clouds, the wet streets or the reflection of water in this time. They return again and again. But the bright fantastical transformation of spring is only brief, and no eye swift enough to capture all its beauty.

You would have to walk through gardens and avenues every day to observe the shimmering green sea which is always similar and yet never the same. Many trees are already entirely clad in rich, deep green and alongside them are bare branches, merely studded with thin, glistening points. They alternate between almost white green, golden-brown and bluish, and snowy blossoms ripple atop thick bushels between them here and there.

There are trees with succulent branches curving upwards in great arcs bearing yellow feathery nests at their tips, others where budding blossoms crouch on the branches like little tents. There are acacias, with their long-grained curving trunks in black and grey, and slender branches proffering tender feathery foliage.

Then there are chestnut trees interwoven with white spots and somewhere on a broad lawn stands a lone chestnut, its crown full and heavy, its blossoms as red as watermelon flesh, its thick cover bright and shining like the pelt of a predator.

The variety of treetops is infinite; some are thick and impenetrable, their spherical forms plainly visible from a distance; others so light and spare that it is impossible to tell whether the branches are advancing toward us or turning away. Only the broad outline is apparent; everything else is just colour, tender and gauzy like sheer fabric. The hazy pale blue sky shines through, so too the luminous white clouds. Then there are crowns that are composed of nothing but little green crests, rhythmic alternations of vale and mount. Others look as though their heavy cover of leaves were sundered, sunk down to the lower branches in great irregular sections leaving deep green hollows where the edges gape and exposing the black trunks.

But the trees don't stay the same, they change as we walk around them, and when the sun is behind us, before us or to the side. A cloud lays grey uncertain shadows over a tree, its colours turn duller, its shadows softer, its contours less defined as the form loses clarity. Sometimes this transformation is unfortunate, sometimes it offers fresh charm as the whole turns wispy and crepuscular. There are no rules, just as there are no rules for those who watch. Naturally a quick eye is an advantage, as the colours and shades change without cease. And because we ourselves are constantly in motion, the transformation is unending. Curiously we often talk of 'nature' as though it were something fixed, unchanging, and then there is Naturalism – the art form that takes nature as it is. But these are not the views of an observer, who knows that nature never holds still long enough to be captured. If we approach a tree, and if the sun stays in place as we do so, and there is no white cloud to cast a sudden new light, it still changes with each step we take. At first we see nothing but the outline, an even, shimmering colour and below it the trunk. We advance, the colours become more vivid, what previously seemed to be just one colour dissolves into spots of various shades, the form changes, the tree extends further upwards the closer we approach, and this towering effect is ever more palpable and impressive. The cloud-like form of the crown dissolves, individual little bushels stand out, the branches we previously held to lay at a single level now extending fore and aft. And now the branches assume new colours; you see the bark of the trunk, the individual leaves. And now it is a completely different entity than the one we saw from a distance.

But even this only describes the bare minimum, and nothing at all of the wondrous unparalleled spectacle of the shimmering giant as it appears to slowly grow above our heads; slowly, slowly its grey limbs become bright shining arms and its illusion of sparsity dissolves into thousands of shining leaves.

We approach an oak tree. Forget all the foolish cant about the oak so proud and bold, words unworthy of a blind singer. Gnarled and wilful is the oak, its leaves curly, bizarre, almost alien, like a tree in a Japanese drawing, like a beast from a far-off land. No matter that it is native to our country. It is only the appearance we are concerned with here. And it appears wondrous, turbid, dry, almost hostile, terrifying, mysterious, a tree of strange and mighty forces. At least that's how it is in summer; now it is altogether different. The tousled tumult of the heavy branches spreads out wide, our tree almost forming a flattened dome with an imperceptible peak, and around the dark branches floats a golden transparent cloak that seems to shine forth from the black branches. As we approach, the branches appear to be hemmed with a blaze of brilliant yellow. And when we get to the trunk, a black broad crown of thorns spreads above us, sending forth a thousand flames, a wondrous glow against the dark blue sky.

A large group of trees, giants standing at wide intervals and spreading their enormous branches; but the outer ones lean more to the side, for the leaves crave the light and those in the centre are better able to stretch out. The grey trunks gleam in the sun, isolated shadows of its branches appear completely black. Together the trees seem to compose a common cover out of the sparse branches, appending them with swarms of leaves, smooth and quite golden. They hang from long, thin, curving stems, moving constantly with the gentle breath of the wind so that they alternate brown and yellow – a rustling, a ceaseless singular shimmering, shining gold like a luminous veil across the blue sky. Then comes the heavy rain, the leaves grow and the golden sea turns green and silver.

POTSDAMER PLATZ IN BERLIN

How many writers have poured out the full measure of their scorn on our cities, how many thousands of times have they denounced them as woefully unhealthy and deadly? This I don't dispute. But it is here that we must live. And there are many places in the wilds of nature that are a thousand times crueller, more unhealthy, more deadly. But there is no use frittering our time away in complaint. Instead, let us try and see just what has become of our *Heimat*. So I wish to describe something of the beauty of a Berlin square, introducing it to those who may encounter it, but also inducing others to view different squares with the same rigour – perhaps they will find similar beauty there, or something completely new.

It is nine o'clock in the evening. The days are warm and dry. The night sky is still bright around this time. The entire square is illuminated by two enormous electrical light masts, spreading their glow like a great dome over the square. The buildings gleam in bright gentle shades, more colourful than in the white light of day. One is slightly reddish, another yellow, and between the buildings lie the thick, craggy mounts of the leafy trees and the crepuscular gloom of the dimly lit sidestreets. The sky is bright green above Bellevuestrasse and Kaffee Josty to the west; the illuminated advertising with its harsh, squalid light, normally so abominable, now shimmers softly in the luminous green. Potsdamer Station recedes in the darkness, the illuminated face of its clock floating like a yellow ball in a sky that is deep blue at this point. The main cornice shimmers in delicate grey, below it a sombre red, almost devoured by the smoky mist that seems to envelop everything. The whole thing gently coalesces, rendering depth almost imperceptible. Contours melt entirely. Perhaps if we fix on a line we might see it distinctly, but only that line; once we absorb the whole image with our eye, everything softens. Black-clad figures form wondrous dark spots on the light grey asphalt. A cab comes toward us all grey on grey, it passes to within twenty paces of us, and it is only your awareness that tells you that the horse's head is closer than the cab, you don't actually see it.

Our perception of depth is in itself a wondrously complex thing. We claim that we see spatially, or physically, as though we could see everything in a space equally. But in reality it is only that which is in the plane before us that we see in its correct proportions. Everything else, everything that advances toward us, or recedes away from us, shifts and distorts in our vision. As children we must learn to comprehend these shifts correctly, and to perceive depth into a flat image as it were. And often we have to observe one and the same object from various points of view. In this way we manage to interpret the distorted image correctly. We are aided in this by the fact of the air changing colour; the further a yellow wall recedes from us, the flatter and greyer its colour. And the further any object is from us, the more it appears to be enveloped in a veil of mist. Now, the brighter it is, the greater the difference in distance we perceive, and the darker it is, the greater the physicality with which we see it, and differences become even less apparent, until all physical vision ceases in the depth of night. But the air quality is also a significant factor – as it becomes hazier, the veil thickens and the distinction between near and far diminishes. Dust, moisture and illumination alter it as well. And just about anyone arriving in an unfamiliar place for the first time will misjudge distances. Because we are not necessarily concerned with seeing things, but understanding their true position and size. In childhood we learn to see true things with their distances in distorted pictures, so much so that we believe ourselves to be seeing them directly. Should we later wish to learn how to see purely for the sake of vision, and the joy we might derive from it, then we must first ignore what we know and what is not actually in the image at all. We may know that the woman's dress there has a razor-sharp edge, but we don't see it. We know that the back wheel of the cab is a long way behind the horse's head, but we don't see it. And now that we wish to simply enjoy the image, it is not a cab at all any more, but rather a wonderful throng of grey patches drifting along over the grey plane of the asphalt.

Dark Potsdamer Strasse – from where we stand the trees conceal many of its lights – issues a dazzling, sparkling, colourful stream of coaches and pedestrians without cease. A mass of shimmering spots that only shine forth under the darkness of the trees, then become larger, then enter the bright light of the square and finally glide past us like large shadows or a bright glow. This play of colours is constantly changing yet always the same, always fresh, always compelling, and yet never fully captured by the eye, like the thousand pearly colours in waves breaking on the seashore.

Weary, we turn around and head for Leipziger Platz, keeping to the old black railing along the side, peering over the green of the large lawn where magnificent ancient trees congregate like gathering clouds, shimmering with the grey-green hoar frost of their blossoms, between them luminous bright facades with dark windows, the heavy blue night-time sky and its stars sparkling above.

Two hours later. Vehicles and people glide past this way and that. We stand at the corner of Bellevuestrasse and look diagonally across the square. The sky is completely black and the darkness seems to enclose the light dome of the square from all sides like a black wall. But this blackness is hazy and smoky. The nearby buildings shine forth brightly from the smoke. Sometimes it so happens that the square is completely empty, and then the grey, smooth asphalt spreads right out into the distant streets, still visible thanks to the glittering lanterns. Above, however, is the night. And so the smooth grey plane seems to hollow out the dark smoky masses of air, as it were, and the lanterns appear to brighten those hollows. Then suddenly everything changes. The whole square before us is covered with coaches. The trams, luminous glass ships, rear up, with the light, thin, tender constructions of the cabs huddling before them. The light comes from a thousand sources, and so even beneath the wagons it is quite bright. On their wheels they seem to float in the air, like enormous insects with unusually segmented bodies and numerous legs. And amid this fairy-tale throng, a wondrous flowing fabric of light shadows spreads over the grey, smooth ground, its outlines alternating clear, soft, light and heavy, or tender, slender and diffuse. A wondrous, restless, errant second world on the ground, bearing no relation to the one above, the one that sired it. The crowd disperses, in a flash the shadows dissolve, and again the bright asphalt spreads out under the luminous dome of light, reaching well into the distance, to the streets filled with smoky darkness, hollowing them out entirely and shimmering in the gleam of their blinking lights.

After the humid days comes the rain, cleansing the air of turbid dust. No longer is the dome that seemed to slice the bright light into the dark black night stretched over the square. The sky is receding, everything is now brighter, the streets wider and deeper, colours strong and bare. And the grey asphalt is dark with rain, thousand-fold reflections mingle in the play of shadows, bright masses of light seem to shimmer below the ground, strangely disjointed from the dull non-reflective parts of the street; somewhere there is a wondrous green strip, reflecting the little signal lamps of a tram.

THE ART OF IMPRESSIONS

Before I talk about the Impressionists at the Artists' Association Exhibition, I would like to try to explain what Impressionism actually is. The same misconceptions arise again and again, leading to the same facile judgements, infantile, ignorant grumbling that rejects the entire movement, most likely on 'nationalist' grounds! Because people do not comprehend the new style, ascribing questionable motives to it and writing off the whole movement as an artistic sham of certain artists, art dealers and Jews, who naturally adulterate the Christian Germanic spirit. This fatuous nonsense compels us to portray the core of the issue objectively, and that is what I aim to do here.

We spend our childhood years forming conceptions of things through the impressions of the eyes, ears, taste buds, and so on. We learn that the many different views of a chair belong to one and the same four-legged object, that it is solid and stable enough to sit on, that you can tip it over when you move around on it too much. The individual impressions of the chair recede behind the gradually acquired conception of 'chair'. And so it is with everything. We begin with the impressions, but from these impressions we always select those that are germane to our lives, and together these partial impressions form our conception of 'chair', 'house', 'school', and so on. When a child then begins to draw objects, problems soon arise. Stairs necessarily form part of our conception of a building, for instance. And when I drew my first building, and had happily committed window and door to paper, I drew the stairs diagonally across the facade and could not understand why this wasn't correct. This is how children and primitive peoples draw. In Steinen's book about his Brazilian expedition you can find fine examples of this. It takes a while until the child understands that you may only draw what you see, not what you know. When the child draws a wagon, the wheels around it are round, because the child knows them to be so, and then it becomes difficult to accommodate all four wheels. These difficulties that we observe in children as they attempt to draw were also encountered by the earliest artists. It was a long time before artists learned to draw the rectangular side of a cube at an oblique, thus achieving the illusion of perspective. The Japanese never managed perspective correctly, and even the Renaissance masters struggled to teach it. And yet our perspective drawings are only a very rough approximation of what we see. For they offer the impression that would come from staring straight ahead with one eye, with the ability to see equally well with every part of the retina. In reality we see with two eyes, and in constant motion, and in a sense we are collating a thousand small images gathered from every possible direction and distance. An image cannot imitate this effect, or at best in simplified form. The perspective representation still attempts to reproduce the formal impression, not the conception, and does so to an infinitely higher degree than primitive perspective or even geometric representation.

But even when the depiction of the form was drawn from the impression, colours were still not painted as seen, but rather as they were known to be. If a coat was blue, it was painted as blue, with no acknowledgement that this blue might not be the same in the shade, or perhaps wasn't even blue at all. The shadows may have been painted a few shades darker, not because this is how shadows really look, but rather because they knew the shadow to be there and sought to reproduce it. It was the Venetians who first discovered that shadows, unilluminated areas have their own colours, colours not seen in the garment when it is held to the light. Then gradually people noticed that the green of the tree was completely different depending on distance and illumination. And they learned that in enclosed rooms the colours become strangely muted and dull, so it wasn't enough to draw the room in perspective, you had to paint it. Rembrandt, for instance, painted the transparent darkness that defines the bright part of a room. Velásquez painted the infinite range of fine grey tones that form in dull interior light.

In this development we can plainly see the effort to depict what was seen, and to entirely forget conceptions of things as they were known, and this effort has become increasingly successful. But it only became an objective in itself in 19th century France, through the French Impressionists, through Manet and Monet, through Renoir and Cézanne. First they managed to forget what they knew, and then to only paint what they saw. They recognised that there is a layer of air over everything, and that 'greater precision' of seeing had destroyed this layer, more or less, along with the unique charm of its beauty.

    We can adjust our eyes to different distances. This is what we call the accommodation of the eye. As we are usually interested in the object rather than the impression, we constantly adjust our eyes, thus shattering the unity of what we see. We jump, as it were, to and fro. Yet the foreground and background are merely representational, they are not what the eye perceives. Previously artists assembled onto the canvas different objects as each was seen through precise observation. So things that were viewed with completely different eye adjustments were brought to a single level where they could only be seen by the viewer from one and the same position. This manner of depiction, so readily proclaimed as natural and self-explanatory was, in purely objective terms, false. This has no bearing on the artistic value, because the objective of art is beauty, not representation. But this destroys a peculiar charm of the visual impression – the layer of air that lies over objects, covering them like a veil, fusing their colours to form a whole. Naturally the eye oriented solely to practical ends sees none of this; its curiosity – the desire to see precise objects – destroys this layer of air. However, once you cast aside this curiosity and see without wishing to see anything in particular, without focusing on anything, you will soon discover all the beautiful things you have missed. I have attempted to depict this here in various ways.

As soon as you decide to capture such beauty in paint, a number of difficulties present themselves. The contours remain blurry. In essence sharp contours exist only for one eye. Together the two eyes see different contours that merge to form one blurred contour. But even with one eye you only see sharp contours when you look closely, thus breaking down the image. This is why the Impressionists show contours as blurred, a little indistinct. Unseeing people who regard the sharp contours of old paintings as their inalienable right fly into inexplicable rage, ranting about laziness and daubs. And the honourable German declares that such dissolution may well be permitted the dubious French, but that it does not befit the loyal German heart. Such good souls would be better advised to open their honourable eyes for once and repent that they so readily wear their hearts on their sleeves. Blurred contours fulfil the highest requirements of perspective representation, they alone manage to depict the layer of air and its charm.

There are other, far greater difficulties. When we see a part of nature in the manner described, and wish to paint it, we simply cannot transfer what we see on to the canvas. This is because the canvas is not within the eye of the beholder, but rather at a distance. In this distance there is air, which modifies and minimises this impact.

Patches of colour that are distinct on the canvas merge for the eye. Therefore we must intensify the colours we see, make them more distinct. Something that is a barely perceptible green shimmer for the eye must be a green spot on the canvas. The light, bluish veil in the shade of a tree becomes a distinct blue patch. If we come too close to the painting we do not understand this blue, or this green. And the man of reason dismisses this as 'crazy'. But as soon as you think about it, you realise that it can't be done any other way. A painting can only be painted for a certain distance, because every distance represents another effect. That's why the observer must be held at a certain remove through the art of painting. That's why you see such coarse patches of colour that only disappear from the canvas at a certain distance to form an image. As you approach, the whole breaks down into incomprehensible spots. The German resorts to furious bluster once again. And those paintings in which every detail is recognisable he praises to the heavens, finding in them industry, fidelity, honesty and a whole garland of civic virtues. And yet these paintings are dishonest. Yes, you can see them from close up and afar, they offer something to every perspective, but they are not wholly satisfying at any remove, as they must always take the others into account. They wish to offer everything and cannot offer anything entire. Impressionism, the painting of impressions, is satisfied with one standpoint, one impression, but offers it entirely and without falsification, with all its beauty. It doesn't paint bodies and the shadows that befall them, but rather both at the same time. So it doesn't matter whether a smudge of paint is a colourful object, or a colourful shadow, or a shimmer created by the air, it doesn't paint the world that we know, the people, the nature, or the intellectual content of which art historians love to speak, rather it paints the wondrous shimmering play of colour that strikes our eyes, regardless of its significance in the rest of our lives. Impressionism is the most complete representation of our impressions that we have found to date, an admirable expedient, the end result of a search centuries long.

But only an expedient! Impressionism is certainly infinitely superior to old methods. But that doesn't make Impressionist paintings better than earlier ones. It is not the method of painting that determines artistic value, rather the beauty that is generated through it. Certainly Impressionists are capable of trivial paintings. On the other hand, there are endless new possibilities in Impressionism that we can barely conceive of today. And therefore we have good reason to distrust painters who shut themselves off from this wealth, desperately clinging to old methods. There can be no denying that some new beauty may be won even there. Hodler is one example. But generally we should expect little from those who have the great fortune of living in an age to which this new wonderland is revealed, and yet fail to stretch their arms out to this wonder.

AFTERWORD

August Endell was far from the only observer to cast a critical eye over the frantic expansion of the metropolis in the early 20th century, but he was certainly the era's most eloquent champion of the city as a source of visual enrichment. And he was also one of the few to back up his theories with interventions in the built environment. As one of the formative figures of Jugendstil – the Germanic equivalent to Art Nouveau – he forms a critical link in the development of new forms in early Modernism.

Unified Germany was just a few weeks old when August Endell was born in Berlin in 1871. Not just a child of empire, August effectively stepped into the family business; his father Karl Friedrich Endell was an architect later appointed the city's head planning director (brother Fritz became a noted illustrator). August's childhood was far from happy; his mother Marie, and two sisters, died while he was still an infant. His intellectual development, at least, benefited from his enrolment in the elite Berlin secondary school, Askanisches Gymnasium. Endell senior died in 1891, leaving August a considerable sum although it disappeared in mysterious circumstances a few years later. Money problems would be a constant throughout his life, along with ill health.

Endell studied philosophy, psychology and mathematics (architecture and design were conspicuous by their absence), first in the historic university town of Tübingen and shifting to Munich in 1892. The city – and in particular the bohemian district of Schwabing to which Endell gravitated – exerted a magnetic attraction on writers, artists and thinkers in Germany and beyond. The benign regency of Bavaria's Prince Luitpold fostered an enlightened haven, with Munich's literary, artistic and lifestyle innovations eclipsing those of Berlin.

Endell's time in the city shaped his intellectual development and set the parameters of his career. In the closing years of the 19th century he concluded his studies, made numerous important contacts, honed his appreciation of art in the city's outstanding museums, began formulating his own aesthetic theory, issued his first publications, and saw his first, and most daring, architectural design become reality – a signature contribution to an emerging new style.

Philosopher Ludwig Klages provides a less than flattering description of Endell around this time as 'lanky, hunched, pale, with a prominent hawk nose, large watery blue eyes, dirty blond hair and a thin, squeaky voice,' concluding that he was a 'stork made man'. Another contemporary described him simply as a 'brain on two matchsticks'. But it was a brain very much attuned to the zeitgeist, and in 1893 we find Endell already enthusing about the Secessionists and other manifestations of the artistic avant-garde. Soon the new arrival was grandly proclaiming the focus of his learning to be 'ethics and aesthetics'. His interests intersected with the teachings of philosopher Theodor Lipps, whose focus on empathy in critical thinking clearly made an impact. But Endell insisted that he was his own man.

This defensive individuality was seen again in 1895 when Endell met the man who embodied modern Munich more than any other – Stefan George. It was a fortunate meeting; the highly influential poet generally shunned personal encounters with all but a coterie of acolytes, preferring to let his mystique operate from a distance. Endell's need for intellectual autonomy notwithstanding, George's advocacy of 'art for art's sake' resonated loudly with the aspiring designer (George, meanwhile, found Endell overly analytical and negative).

Yet another of the motifs in Endell's life emerges around this time – his association with strong, self-assured women who boldly rejected the Wilhelmine feminine ideal. Later this pattern will introduce us to Else Plötz, Lou Andreas-Salomé, Franziska zu Reventlow and Lily Braun. But before them came two women whose friendship with Endell had far-reaching consequences for his career. With short hair and distinctive reformist clothes, Anita Augspurg and Sophia Goudstikker lived in relative openness as a couple and established their own photographic studio, the first solely female-owned business in Germany. Located on Von-der-Tann-Strasse near the southern edge of the Englischer Garten, the

Hofatelier Elvira was also a meeting place for allies who shared their belief in feminism and other progressive ideals. Endell, a supporter of female emancipation, soon became a fixture of their cultural gatherings.

It was through Augspurg and Goudstikker that Endell made another crucial contact in 1896 – Swiss-born designer and sculptor Hermann Obrist. Recognising Endell as an original thinker anxious to make his mark, Obrist encouraged him to not only write about but also practice architecture and applied arts – despite his lack of formal training in these areas. Endell and Obrist would become key figures in a style that drew from and paralleled Art Nouveau, which was then emerging in France. It was inspired by the sensuality of natural forms and aimed at combining harmonious elements to form a cohesive, enveloping whole. It was an aesthetic approach reflected in the journal *Jugend* which was founded later that year, and which would lend its name to the movement. Jugendstil genuinely was an *art nouveau*, a new art, one that rejected the pathos and pastiche of state-sanctioned historicism. The radical subordination of form to function would only take hold on the far side of the First World War, but for a moment, Jugendstil satisfied a hunger for design worthy of the coming century.

Endell's first book was a volume of poems produced privately in just 50 copies entitled *Ein Werden* (A Becoming), issued in 1896. But his breakthrough as a theorist came later that year with the slim publication *Um die Schönheit* (On Beauty). It carried his Rorschach-style image of an orchid on its cover and the subheading

'A Paraphrase of the Munich Art Exhibition 1896'. The text is prefaced with a verse by Charles Baudelaire; the *grand-père* of Modernism was a bold reference point for one currently engaged with establishing himself as a poet, an art critic and a herald of new forms. Half of the text is taken up by Endell's digressive thoughts on unprejudiced observation, the role of art, and the emotions that it might awaken, before he turns to the exhibition that is the book's putative subject. He evinces an almost total disinterest in the 'meaning' or 'message' of art, instead proposing a radical, almost mystical aestheticism, a fundamentalism of form. It was an approach that called less for a critical eye than a willingness to submit:

> But those who learn to give in to their visual impressions completely, without associations, without secondary objects of any kind, those who just once feel the emotional impact of forms and colours, will find them to be an inexhaustible source of extraordinary, unimagined pleasure. And the moment when the understanding for these things first awakens should be an event in every person's life. It is like an intoxication, a madness that comes over us. The joy threatens to destroy us, the profusion of beauty to suffocate us. Those who have not experienced this will never understand visual art.

It is a spirited text – combative, self-confident, extremely opinionated. Among those intrigued by Endell's words were novelist and pioneering psychoanalyst Lou

Andreas-Salomé. She sought Endell out after reading *On Beauty* and would prove a life-long friend. It is highly unlikely that Endell – captivated by Friedrich Nietzsche like so many intellectuals of his generation – was unaware of Andreas-Salomé's connection to the philosopher, who was still alive yet permanently incapacitated. Although ten years older than Endell, the Russian-born Andreas-Salomé was just as alert to the changing disposition of the times. An extraordinary photo taken in a summer house outside Munich in 1897 shows the two with Andreas-Salomé's then lover Rainer Maria Rilke, along with writer Frieda von Bülow and Russian critic Akim Volynsky. In a single image we see the changing status of women, the new spirit of internationalism, the breaking down of social castes, all delivered with insouciant modernity.

But Endell wasn't content with mere modernity; he was now forecasting the future. In articles for journals like *Dekorative Kunst*, Endell presented the case for abstraction – the 'art of forms' as he called it, 'which depicts and symbolises nothing, which functions through freely invented forms, just as music does with free tones.' Music was a recurring point of reference; in 1898 Endell claimed that he and his contemporaries were 'not just at the start of a new stylistic period, but also beginning to develop an entirely new art … that stirs our souls as deeply, as strongly, as only music can in the form of sound. The barbarian hates our music.'

Finally Endell made his first major step from theory to practice, and once again his friendship with Augspurg and Goudstikker was key. They planned to renovate their studio and entrusted the work to Endell. To their lasting credit they approved his revolutionary designs, which one official claimed 'made a mockery of the art of drafting'. Endell's design was a *gesamtkunstwerk*, which he insisted would incorporate 'the entire interior decoration, doors, window friezes, capitals, cornices, painted ornamentation, banisters, grates, wrought iron doors …'. But it was the extraordinary exterior that drew the most attention.

With little requirement for natural light, the largely windowless facade offered a blank canvas. Endell filled it with a flourishing squall of a motif that has variously been described as a dragon, a wave or – in the words of art historian Erich Franz – 'the first abstract work in art history', which would certainly fit Endell's theories of the time. It shock value was undeniable; the workmen tasked with realising Endell's vision avoided the local taverns, fearing the ridicule of other labourers. Opening just weeks ahead of the 20th century, the new Hofatelier Elvira was the boldest expression of Jugendstil to date, a *succès de scandale* which made Endell's name. Architect and town planner Johannes Otzen was just one of the old guard appalled by these new forms, seeing in them the dread influence of Nietzsche, and making dark references to 'Übermensch-Ornament'.

Conversely Endell's *actual* Nietzschean designs were far more conventional. With the approval of Elisabeth Nietzsche, who did so much to distort the reception of her brother's work, Endell designed two volumes of works – *Also sprach Zarathustra* and a collection of verse and aphorisms. Both bore the motif of a snake forming a loop, symbolising the Nietzschean 'eternal recurrence' that Endell would later reference in his own work.

In the first spring of the new century, August Endell's life was upended by the arrival of actress, poet and aspiring art student Else Plötz. Ludwig Klages maliciously dismissed her as 'a Berlin courtesan who had passed through many a hand in intellectual circles'. From a more enlightened perspective we might say that her pursuit of male partners embodied a conception of female erotic self-determination that was decades ahead of its time. Plötz had approached Endell for art lessons, but their association soon turned amorous.

In 1901, the pair moved to Berlin. While Munich was still the country's pre-eminent cultural centre, signs of the coming shift to Berlin were already manifesting themselves. Endell joined the likes of architect Peter Behrens, who saw greater opportunities in the rapidly expanding imperial capital. As if in confirmation of his instincts, Endell soon received his first major Berlin commission: the Buntes Theater, a new venue for the city's cabaret culture, then in its infancy. It offered a significantly larger canvas than the Elvira, and the result was a lively expression of Jugendstil ideals, fecund with vegetal forms.

Endell and Plötz married that same year – she, typically, had proposed to him. The pair brought out each other's eccentricities, striding the streets of Berlin in unusual clothes of Endell's devising and giving each other Chinese names; Endell was 'Tse' and Else 'Ti', 'master' and 'yellow' (representing majesty) respectively.

But it was an ill-starred marriage. Else found August sexually unsatisfying and their union soon expanded to include her new lover, writer Felix Paul Greve, who had already distinguished himself in Munich as a fraud and a fantasist.

In 1903, the curious threesome travelled to Italy, although the third wheel came loose when Endell, overworked and wrung out by his disintegrating domestic circumstances, attempted suicide in Naples. He made his own way to Ravello where he caught up with two old Munich companions, philosopher Karl Wolfskehl and writer Franziska zu Reventlow.

His inevitable split with Else formalised in 1904, Endell returned to his long-term programme of broadcasting his aesthetic ideals. He opened his own design school and the untaught creator became a teacher. He continued to write, with a series of articles for the journal *Freistatt* covering a range of subjects including art education, Vincent van Gogh and the newly opened Protestant cathedral in Berlin. Meanwhile his ex-wife was contributing erotic verse to the same journal under the name Fanny Essler, which confusingly was also the title of a book by Greve that drew a thin veil of fiction over Plötz's life, in which the hapless Endell appears as 'Eduard Barrel'.

Endell soon found another forum for public enlightenment. *Die Neue Gesellschaft* (The New Society) was a new 'socialist weekly' founded by the husband-and-wife team of Lily and Heinrich Braun. Their British equivalents, Beatrice and Sidney Webb, contributed to the first issue in April 1905, alongside a piece by Kurt

Eisner, later head of the short-lived 'Räterepublik' in post-First World War Bavaria.

Endell's contribution to this issue is entitled 'Art and the People' and – ever the contrarian – he leads with the statement 'I am not a social democrat'. But he also goes on to say 'I am convinced that a profound artistic culture is only possible when there is a vital and mature desire, and appreciation, for art among the working class'. The ten articles that Endell submitted throughout 1905 range over topics that he would elaborate in *The Beauty of the Metropolis*. Of these, Endell would later specifically nominate six as forming a complement to that book, which are included in this edition.

The second article, 'On Vision', is something of an aesthetic manifesto, opening with the bold declaration: 'The objective of all the arts is beauty'. Endell aims to liberate vision from forethought, returning to the 'surrender' of *On Beauty*, but also reaffirming the democratic potential of this seeing, 'something that can be practised anywhere, at any time', concluding that 'today, the present, this reality is the most fantastical and incredible thing of all'. In 'Spring Trees', one of his most rhapsodic pieces, Endell tenderly follows the development of the new season's growth. He returns to these observations in his next contribution, 'Treetops', as if reporting on a sensational trial of which the public was avid for the latest news, rather than a natural process that recurs every year. The minutely observed alertness to plant life that informed his ornamentation for the Hofatelier Elvira and the Buntes Theater is very much apparent here.

These two pieces were bracketed by sketches of public spaces in the German capital – 'Evening Colours' and 'Potsdamer Platz in Berlin', in which Endell exhibits a photographer's sensitivity to changing light. These highly impressionistic accounts lead organically into three pieces that return to the art appreciation of *On Beauty* – 'The Association of German Artists' Exhibition in Berlin', 'The Art of Impressions' and 'Our Impressionists'. These were inspired by that year's instalment of an annual exhibition held in a long-gone hall near what is now Berlin's main railway station.

These three articles are dominated by capital 'I' Impressionism. There is a continuum here; an Endell study of a Berlin street could just as easily be a description of an urban scene by Monet. Impressionism – the alertness to and recreation of fleeting moments, depicted as they appear and not as convention dictates they *should* appear – was central to Endell's aesthetic vision. The origins of Impressionism were older than Endell himself, and by 1905 had long extended beyond France's borders, with local painters such as Max Liebermann and Lovis Corinth proving early adopters of the new form. Endell's rugged defence of the style may suggest that he is restaging a culture war whose outcome had already been decided. But while the early rejection of Impression is part of the mystique of its reception, its later reproduced ubiquity obscures how *sustained* this resistance was. A generation after it first appeared, Impressionism continued to meet with stubborn opposition from Germany's art establishment, which was able to discern the imminent downfall of civilisation in what we might see as a pleasant study of the bourgeoisie enjoying a nice day out.

There is an argument for positioning Endell's acutely observed texts within the far less heralded tradition of Impressionist literature, which in Germany would include the poetry of Detlev von Liliencron, and to an extent Rainer Maria Rilke, and the fiction of Eduard von Keyserling. There is also a parallel with the even earlier 'Sekundenstil', one of the innovations of Naturalism – in which imagined events of short duration were described in 'real-time' detail. All of this, it has to be said, was flighty fare for a serious-minded journal of current affairs and it is only with his last contribution, 'Workers' Houses', which addressed contemporary housing schemes, that Endell actually offers the kind of article one might expect of an architect writing for such a paper.

In 1906 Endell received another major commission in Berlin. The new, centrally located Hackesche Höfe complex made a virtue of the deep courtyards that were a feature of many Berlin buildings at the time. With mixed residential, commercial and even light industrial usage, these spaces were often dark, dingy and chaotic. The Hackesche Höfe were something completely different. The first and most prominent courtyard, in Endell's design, was an oasis of serenity, with a generous open area, large windows, and glossy tile work in predominantly pale shades making the most of the available light. While there is a distinctive rhythmic élan to Endell's patterns, it is far more restrained and geometric than his previous ornamentation. The commission also included the design of a varieté theatre, while the following year brought a major residential project in Charlottenburg, the Haus am Steinplatz.

Endell was entirely a creature of the city. It was his home, his place of learning, the showcase for most of his works, and had already figured prominently in his writing. His minority had coincided with an astonishing population boom in Germany, and the growth of urban centres in particular. At his birth, just five per cent of Germans lived in cities; by the time he reached adulthood city-dwellers represented half of the country's population. Endell had experienced first-hand a process of accelerated urbanisation unparalleled in European history to that point.

All of this found expression in Endell's most enduring text. *Die Schönheit der grossen Stadt* (The Beauty of the Metropolis) appeared under the editorial guidance of Wolfgang von Oettingen, a Baltic German professor of literature and fine arts, and was published by Strecker & Schröder in Stuttgart in 1908. That same year Endell became engaged to sculptor Anna Meyn, and perhaps this new romantic chapter helps explain the tone of rapture that recurs in the book.

But Endell has a lot to say before he turns to his main subject, the city. Here we see again the pugnacious commentator of *On Beauty* and the *Neue Gesellschaft*, not to mention the man who had taken on the nation's highest (self-appointed) artistic authority. In 1901 Wilhelm II had given a notorious speech at the opening of a vainglorious outdoor display of his ancestors and other overlords in marble effigy (where historical evidence was lacking, some early rulers were modelled on the Kaiser's hangers-on). His address decried 'gutter art' – understood to mean all modern art and literature. All that was good and fine and admirable in art, insisted the Kaiser, had already been achieved, and the creative professional's task was simply to re-assemble these fixed, historical forms in different ways. This was precisely the kind of vapid historicism that infuriated Endell. In 1902 he addressed the Kaiser directly, in Maximilian Harden's anti-establishment weekly *Die Zukunft*. He mocked the 'love of fatherland, love of the military, devotion to hereditary rulers' that Wilhelm represented, and picked up on the Kaiser's contemptuous reference to the 'gutter'. 'But is it really a sin,' asked Endell, 'to show the poor people who spend their life in the gutter, that even there in the most hideous corners of the city, they can find beauty, beauty that may give them the strength to overcome misery and torment?'

Here we find the core sentiment of *The Beauty of the Metropolis*. The preamble, which consumes a fifth of the text, criticises those who seek refuge in the past, particularly in historicism – a Freudian might have noted that this was precisely the style in which Endell senior had worked (and that August describes the city as a 'mother'). He is no more warmly disposed to the kind of lifestyle reformers who were preaching a 'return to nature' at the time. Endell takes a further swing at Romanticism, then undergoing a revival, and the reflexive denunciation of materialism. He launches a bucket of cold water at utopian idealism, grandly pronouncing that 'The objective of humanity, whether it accepts it or not, is acquisition.' It is something you can almost imagine Ayn Rand saying at a later date.

But Endell was never an easy man to pin down ideologically. He takes on a key right-wing shibboleth, fundamentally revising the concept of *Heimat* which is key to his exploration of place. Frequently yet incompletely translated as 'homeland', it is a word with extensive emotional resonance in German, tied to memory and ancestry. It was generally attached to the rural region of one's parents, grandparents and forebears. The literary form of the *Heimatroman*, which emerged around 1890, favoured the rural idyll, traditional morality and accustomed gender roles over the metropolis, progressive values and independent women. To the reactionary nationalist sentiment that often dwelt within the *Heimatroman*, the city stood for the unholy trinity of socialism, modernity and Jewry.

In *The Beauty of the Metropolis*, Endell suggests that *Heimat* is not just an automatic inheritance but one shaped by experience and sensibility, something mercurial and highly subjective. As he warms up to his main theme, Endell decouples *Heimat* from the patriotism with which it had become entwined, at once rejecting national chauvinism and suggesting that in fact your *Heimat* could just as well be a city.

As well as sharing his detailed observations, Endell also describes the provenance and construction of the critical apparatus through which he views the city; one paragraph is a survey of art history that progresses briskly from cave paintings to Vincent van Gogh. And 'the city' that Endell references here is usually Berlin, specifically its modern quarters. Although riddled with misgivings about contemporary development in the capital, Endell praises it as a wellspring of aesthetic experience. Where others had found Babylon on the Spree, he finds a 'fairy tale', a 'wonderland', his lyrical flights often transposing his visual impressions into something approaching music. Much of the second half of the text is taken up with a tangential tour of Berlin that begins with the view from his study window and takes in busy squares and railway architecture, sporting grounds and construction sites, female fashions and newly built churches, before finally arriving at the very heart of the city, the Brandenburg Gate, in time for the magic hour.

The Beauty of the Metropolis appeared the same year as Edmund Edel's *Neu-Berlin* (New Berlin), the last of the 51-volume series *Metropolis Documents*, edited by Hans Ostwald. While Edel shows far more interest in social interactions, he shares Endell's sense of wonder at the ever-changing visual spectacle of the modern city. Edel's prose sketch of Potsdamer Platz is remarkably close to Endell's impressions of the same square in both the *Neue Gesellschaft* and *The Beauty of the Metropolis*:

> The night begins. The shimmering haze of the vibrating day dissolves in the half-light of the dusk, individual electric flames play, forerunners of the great nocturnal illumination through the matt blue air. The mighty stands of trees on Leipziger Platz thicken into great masses, the old monumental

> worthies on their pedestals disappear along with
> their contours for some sorely needed rest.
>
> Evening! Peace on Earth and pandemonium on Potsdamer Platz.
>
> The lamps flash – warm, violet-brown light prances from the tall lights … The electric trams buzz, the automobiles huff, the coachmen curse …

Impressionism was still a key touchpoint for Endell. In a major study of the movement published in 1907, art historian Richard Hamann describes it as 'the end-point of development', and this appears to be Endell's view as well, despite his prophetic call for abstraction. His frequent mentions of 'veils' recall the 'atmospheric veil' referenced repeatedly in the literature on Impressionism. But often the 'veils' of which Endell speaks are nothing more than pollution, the effusions from household chimneys and factory smoke stacks that enveloped Berlin's streets. We would have to imagine a present-day Chinese writer eulogising the yellow, choking smog of Beijing to recreate the effect that this would have had on the contemporary reader.

There is idealism here, but naivety as well. Endell's blithe suggestion that the reader might drop into a factory and ask the workers there to explain the different machines and their din strongly suggests that he had not actually attempted this himself. His aesthetic fundamentalism is if anything even more rigorous here, sometimes reaching beyond the bounds of taste. The most troubling example is the 'beauty' that Endell finds in the sunken cheeks of an undernourished child. He appears no more inclined to ponder the 'meaning' of this image than the paintings in which he sees nothing more than a play of colour and form.

Similarly, it is the 'beauty', rather than the dignity of labour that Endell extolls. When he encounters a crowd it is never distinguished by its social make-up or the individual characteristics of its members. But here again there is a democratisation of sorts, by which each person is free to look upon or contribute to the fantasia enlivening the street. And Endell does genuinely appear to believe that anyone might be able to see the way he sees. In the last of the books that Endell issued in his lifetime he is intoxicated by his own observations, dazzled by his glimpse of a new world from which he returns as a visionary of vision.

Endell continued to write for specialist journals and furthered his design work, with a series of eclectic commissions in Berlin including a shoe shop, a racecourse and a series of villas in the well-to-do outlying district of Westend. Jugendstil was falling out of fashion, and Endell's work showed signs of a new decorative continence. In 1912 Endell joined the Deutscher Werkbund, an association that brought together architects, designers and related professionals (and still exists to this day). Here, arguably for the last time, Endell found himself at the forefront. The group – which included Peter Behrens, Ludwig Mies van der Rohe and Harry Graf Kessler – aimed to create physical forms worthy of the new machine age.

In 1914, Werkbund members were split

on the issue of developing a group style. Endell, characteristically, insisted on his independence, siding with the 'individualists' of the group, including Walter Gropius, Bruno Taut and Henry van de Velde. These concerns were soon overtaken by the outbreak of the First World War. While Endell's perilous health kept him from active service, the conflict nonetheless dominated his work in the period, with designs including a war cemetery, a war museum and a home for returning veterans, none of which were realised.

The final episode of Endell's life began shortly before the end of the war, when he was appointed director of the National Academy of Art and Applied Art in Breslau (present-day Wrocław, Poland). While he was valued as a teacher, and the position brought welcome stability, the provincial centre was hardly a hub of progressive culture. He might have been heading the more prestigious art school in Weimar had he not been beaten to the position by Walter Gropius. There, in 1919, as the city hosted the constitutional conference that would gave its name to the new republic, Gropius was taking one of the most momentous steps in the Modernist adventure by establishing the Bauhaus.

Where Endell had questioned the form of ornamentation, others were now questioning its existence. The man who had called for abstract art before artists themselves was now at a remove from the zeitgeist. His dogged attachment to 'beauty' – the word that featured in the titles of his two major book publications, a quality he still claimed as the highest objective of the arts – now looked quaint. Endell was beginning to seem like a dried-up hothouse bloom pressed between the pages of a chapter in design history long closed.

Endell's health, never robust, deteriorated sharply in 1923. Around this time his first wife reappeared in Berlin. She had left for the US in 1909 with Greve, who faked his own death and reinvented himself as Frederick Philip Grove, later to become a major figure of Canadian letters. Else had undergone her own radical transformation, remarrying, collecting a title and making a highly idiosyncratic mark in the art world as Elsa von Freytag-Loringhoven, the 'Dada Baroness'. She now returned bitter and impoverished. She wrote a scurrilous poem, quite possibly an attempt to extort money, in which her ex-husband appeared as 'August Puckellonder'. But this latest indignity was at least of short duration; returning from a holiday in Bavaria in 1924, Endell had a heart attack. He died in Berlin the following year, barely 54 years old.

What remains of August Endell – his words, his work, his world? Certainly his physical legacy has suffered. The Nazis removed the loathed ornamentation from the Elvira studio in 1937, reputedly on the orders of Hitler himself, and British bombers finished the job in 1944. The Buntes Theater was another war casualty, along with the Salamander shoe shop and numerous other interiors. Even Endell's 'resting' place was destroyed in the Second World War. And it wasn't just the war; one of his villas was levelled in the 1970s and as late as 1990 the sanatorium he built on the North Sea island of Föhr was largely demolished. But happily some of his works endured, including numerous pieces of furniture. So too the Hackesche Höfe complex, which was treated to a sensitive post-reunification clean-up and remains highly popular with visitors to Berlin. A number of private villas remain, while the renovated Haus am Steinplatz now functions as a luxury hotel.

Endell's posthumous reception has been similarly mixed. The first anthology of his texts was a curious volume issued in 1928 by a publisher specialising in gardens. Edited by his widow and adorned with a photo of a pink bloom, it sought to confine Endell to the role of 'observer of nature'. Naturally there was no chance of rehabilitation for a hated progressive during the ensuing Nazi era – the barbarian was still deaf to his music – but even in 1955, Anna Endell was complaining to architect Hans Scharoun that there was still no authoritative register of her late husband's work. The 1960s and '70s brought renewed interest in Art Nouveau and Jugendstil, and in 1977, over half a century after Endell's death, a Munich exhibition reintroduced the public to his design work. This also marked the return of Endell the theorist, with the catalogue for the exhibition reproducing the entire text of *The Beauty of the Metropolis*. There were further annotated anthologies of his texts in 1995 and 2008, while a major international conference on Endell in Berlin in 2010 was followed in 2012 by an exhaustive monograph and the first exhibition of his work in his home city. And as this edition goes to print an exhibition in Berlin's Ephraim-Palais is taking *The Beauty of the Metropolis* as the starting point for a survey of paintings depicting urban scenes.

In the annals of design history Endell remains bound to Jugendstil, and his crucial yet transitional position in Modernism finds its most apt symbol in the traffic depicted in *The Beauty of the Metropolis* – hectic, yet still largely horse-drawn. But viewed as a whole, his career appears arrestingly modern. In his outspoken views and interdisciplinary activities – architecture, interiors, furniture, book covers, illustration, fabric, teaching and publishing – Endell embodies a highly contemporary idea of the opinionated design polymath.

Even more relevant for us today is his revelation of the observable riches all around us; from out of the past he offers us a passionate defence of the here and now. Endell emerges as nothing less than a prophet of the individual aesthetic odyssey. Photography is curiously absent from Endell's writing, but it is arguably here that his ideas are most fully realised. If he returned today, Endell the autodidact would no doubt be impressed by the visual sophistication of the average citizen and our ready access to means of image production. Where our medieval peasant ancestors might have been overawed by a once-in-a-lifetime visit to a cathedral, the city dwellers of a secular age daily testify to Endell's insight that splendour awaits the eye wherever it may fall:

> Open your eyes, don't invent miracles or another world above the clouds; for here in your world you have the kingdom of Heaven.